Springer Tracts in Modern Physics 112

Springer Tracts in Modern Physics

* denotes a volume which contains a Classified Index starting from Volume 36

Christian Kiesling

Tests of the
Standard Theory of
Electroweak Interactions

With 86 Figures

Springer-Verlag Berlin Heidelberg GmbH

Dr. Christian Kiesling

Max-Planck-Institut für Physik und Astrophysik
Werner-Heisenberg-Institut für Physik, Föhringer Ring 6
D-8000 München 40, Fed. Rep. of Germany

Manuscripts for publication should be addressed to:

Gerhard Höhler

Institut für Theoretische Kernphysik der Universität Karlsruhe
Postfach 6980, D-7500 Karlsruhe 1, Fed. Rep. of Germany

Proofs and all correspondence concerning papers in the process of publication should be addressed to:

Ernst A. Niekisch

Haubourdinstraße 6, D-5170 Jülich 1, Fed. Rep. of Germany

ISBN 978-3-662-15126-6   ISBN 978-3-540-47445-6 (eBook)
DOI 10.1007/978-3-540-47445-6

Originally published by Springer-Verlag Berlin Heidelberg New York in 1988.

Softcover reprint of the hardcover 1st edition 1988

2157/3150-543210 – Printed on acid-free paper

Preface

Tests of the standard theory of electroweak interactions with recent experimental data on weak neutral currents are reviewed. The data, which range from low energy phenomena such as atomic parity violation to the discovery of the heavy $W^\pm$ and Z^0 bosons, impressively support the theory as formulated by Glashow, Salam and Weinberg.

The available experimental information is presented with emphasis on electroweak interference effects as observed in e^+e^- experiments at PEP and PETRA. These storage rings have completed their scientific programme and physics analyses have been finalised. The observable quantities are the total cross section, the charge asymmetry and, in the particular case of the τ lepton, the polarisation of the final state fermions. Combining the measurements of the leptonic final states with results from νe scattering leads to a unique solution for the couplings of the leptons to the weak neutral current, consistent with the standard theory. Furthermore, results are described on the weak couplings of heavy quarks produced in e^+e^- annihilations which complement the general picture of the weak neutral current couplings to the elementary fermions.

Some new results from ν scattering and, for completeness, charged lepton quark scattering and atomic parity violation experiments are briefly reviewed. These data, together with the results from e^+e^- yield a consistent picture of the couplings for quarks and leptons to the weak neutral current and are fully accounted for by the standard theory. The major triumph of the theory was the discovery of the heavy bosons $W^\pm$ and Z^0 in the $\bar{p}p$ experiments at CERN.

All experimental data can be described by one additional parameter, the Weinberg angle, usually given as $\sin^2\theta_W$. Measurements of this angle by the various experiments yield impressive agreement. Radiative corrections to the measurements tend to improve the agreement and suggest the correctness of the theory at the quantum level.

Despite its success, the standard theory has some deficiencies. In particular, it does not address the family problem which introduces additional free parameters such as the fermion masses and the number of lepton/quark generations. Furthermore the elusive Higgs is far from being understood. Some alternatives to the standard theory are therefore discussed such as models with larger symmetry groups or composite models, as well as further tests of the theory using the next generation of accelerators.

München, September 1988 *Christian Kiesling*

Acknowledgements

Reviewing experimental results on tests of the standard theory of electroweak interactions is, in the present stage of such intense investigations, a never-ending task. To undertake such an enterprise and finish within a reasonable amount of time needs a lot of help from different sides.

My thanks go first of all to my experimental collegues who have provided their data to me, frequently prior to publication, and explained to me many details of their analyses. Among the many contributers I would like to acknowledge especially the cooperation of Drs. C. Mana, H. U. Martyn, R. Prepost, V. Lueth, H. Ogren, B. Naroska, and J. Meyer. I have benefited from numerous enlightening discussions with Drs. A. Buras, J. Kühn, R. Kotthaus, S. Jadach, W. Wittek and Profs. N. Schmitz and F. Wagner. Their helpful suggestions and criticism are highly appreciated. Furthermore I like to thank my collegues in the CELLO Collaboration and at the Max-Planck-Institut für Physik und Astrophysik for many fruitful discussions while preparing the review.

My special thanks go to Prof. G. Buschhorn and Prof. U. Meyer-Berkhout for their interest, encouragement, and continued support of this work.

Expert advice from Prof. P. Breitenlohner about the subtleties of type setting, the skillful typing of the manuscript by Mrs. E. Loskutow and the careful drawing of the figures by Mrs. R. Heininger are gratefully acknowledged.

Finally I would like to thank my wife Ilona and my little sons Kevin and Robin for allowing me to invest quite a number of evenings and weekends in this report.

This work was supported by the Deutsche Forschungsgemeinschaft.

Contents

1. Introduction

Unifying the four known forces between elementary particles, i.e. gravitation, weak, electromagnetic and strong interactions, certainly is one of the most intriguing and demanding goals in physics. The belief that local gauge invariance plays an essential role in the correct mathematical description of particle interactions has proven an extremely useful guide towards that goal. Glashow (1961), Weinberg (1967), Salam (1968) and others (Glashow, Iliopoulos and Maiani, 1970) have formulated a gauge theory based on the symmetry group $SU(2) \times U(1)$ unifying electromagnetic and weak interactions of leptons and quarks which has met spectacular success in describing a large amount of a priori unrelated data and predicting new and unexpected phenomena. By now widely accepted as the "Standard Theory" of electroweak interactions between elementary leptons and quarks grouped in three families, its most important predictions were the existence of the charmed quark and, after the discovery of the bottom quark, the existence of the top quark, the heavy gauge bosons $W^{\pm}$, and the Z^0, giving rise to neutral weak currents. Neutral currents and charm were indeed discovered more than a decade ago, and the existence of the gauge bosons could recently be established with the $\bar{p}p$ collider at CERN. The top quark still awaits discovery, its mass probably being too large to have been produced with a substantial rate at present accelerator energies. The theory furthermore predicts, in its simplest form, a scalar neutral particle, the Higgs boson.

When experimental results in support of the standard theory were still lacking, its theoretical success already provided additional motivation to apply the gauge principle to theories of the strong interaction. As a result quantum chromodynamics (Fritzsch, Gell-Mann and Leutwyler, 1973; Politzer, 1974) built on the symmetry group $SU(3)$ of colour has been formulated, which leads to new gauge bosons, the gluons, and both qualitatively and quantitatively explains many phenomena in hadronic systems. It seemed natural to try to include the strong and electroweak interactions into "Grand Unified Theories" or GUT (Georgi and Glashow, 1974; Georgi et al., 1974). Such theories necessarily couple quarks to leptons and thus generally predict a finite lifetime for the proton. Furthermore, in the framework of the hot big bang model, GUT lead, e.g., to fascinating new insight into the development of the early universe. These concepts of a grand unification of the basic interactions mentioned above, however, have not yet matured sufficiently to be challenged by detailed experimental study.

With the discovery of weak neutral currents in 1973 an intense period of searching and investigating these new phenomena was started in almost all possible reactions between leptons and/or quarks. Most of the experimental results on weak neutral currents have initially come from neutrino scattering on electrons and hadrons. In 1978 an important step was taken with the discovery of parity violation in electron nucleon scattering generated by the interference of the weak neutral current with the electromagnetic current (electroweak interference). A beautiful series of experiments on deep-inelastic scattering of polarised electrons on deuterons (Prescott et al., 1978, 1979) and on atomic spectroscopy (Barkov and Zolotorev, 1978, 1979) clearly demonstrated parity violation and furthermore excluded a large variety of other candidate models. Combining the deep-inelastic scattering data with the neutrino data, a unique solution for the couplings of electrons and quarks to the weak neutral current can be obtained, consistent with the standard theory.

Since then the bulk of new experimental information has come from e^+e^- collisions where electroweak interference effects can be studied at very high momentum transfers (now reaching a Q^2 up to about 3000 GeV^2 at the storage ring TRISTAN). Awaiting the data from the new generation of e^+e^- machines SLC and LEP, the material presented in this report will summarize our knowledge from the high energy e^+e^- experiments at PEP and PETRA, including also the recent results from TRISTAN. The major part of this review deals with a critical analysis of the data from these experiments. In essence, the data consist of measurements of the total cross section and the charge asymmetry in fermion-antifermion production from e^+e^- annihilation. The fermion in the final state may be either a lepton (μ, τ) or any kind of quark, including the heavy quarks c and b.

The experimental material relevant to the weak neutral current other than e^+e^- data has been reviewed extensively in the past (see e.g. Sciulli, 1979; Kim et al., 1981; Myatt, 1982; Pullia, 1984; Klein 1985). The present report also tries to cover the recent experiments not contained in these works, where weak neutral current phenomena play a dominant role: New data from νe scattering, deep-inelastic ν hadron scattering and atomic parity violation have become available and provide, when combined with e^+e^- results, a consistent picture, within the framework of the standard theory, for the weak neutral couplings of the leptons and quarks of the three known generations. Several related topics, however, have not or rather briefly been treated, such as nuclear parity violation, flavour-changing neutral currents and "missing particle" searches. For details on these processes, one may consult the articles of Fortson and Lewis (1984), Ellis (1983), Yamada (1983). Furthermore the entire sector of charged weak current phenomena involving the three generations of quarks is not covered here and may be found, e.g., in the reviews by Pullia (1984) and Barbiellini and Santoni (1986).

The review is organized as follows: Chapter 2 gives a brief introduction to the history, motivation and formalism of the standard theory of electroweak interactions with emphasis on the physical ingredients of the theory. Chapter 3 presents the results from the e^+e^- storage rings, mainly PEP and PETRA, both for leptonic and quark final states. The individual experiments are briefly described and, in more detail, the measurements sensitive to weak neutral current contributions. A new analysis of all available e^+e^- data is presented which tries to improve on the determination of the relevant parameters of the standard theory from these data. In Chap. 4, experiments on neutrino-induced reactions are covered, which lead to a unique solution for the chiral couplings of the quarks from the first generation and to the well-known twofold ambiguity of the weak neutral couplings for the electron which is resolved with the help of the e^+e^- data. Chapter 5 reviews charged-lepton quark scattering, providing additional information on the weak couplings of electron and muon. The experimental discovery of the weak bosons $W^\pm$ and Z^0 and the recently improved measurements of their properties are reviewed in Chap. 6. Chapter 7 summarizes the results of the existing experiments on the weak neutral couplings for the three fermion generations and presents a critical comparison of these results with the predictions of the standard theory. In Chap. 8 some alternatives to the standard theory are discussed together with their possible consequences for future accelerator experiments.

2. The Standard Theory

2.1 Historical Preliminaries

The weak interaction departs in a peculiar way from all other known forces in
Nature: It is not invariant with respect to space coordinate inversions or mirror
reflections, i.e. the weak interaction violates parity conservation. It took more
than twenty years after Fermi conjectured a theory of nuclear β decay, based
on two directly coupled vector currents (Fermi, 1934), to realise that parity was
not conserved in weak decays. T.D. Lee and C.N. Yang (1956) initially proposed
parity violation to explain peculiarities in the weak decays of certain charged
particles, known as the $\theta - \tau$ puzzle: Two mesons, θ and τ, with identical masses
and quantum numbers, decayed into final states containing two (even parity)
and three (odd parity) pions, respectively. Lee and Yang identified θ and τ
decays as merely two different decay modes of the same particle, now known as
the charged K meson, and consequently postulated parity violation in the weak
decay of this meson. Soon thereafter parity violation was established in nuclear
β decay (Wu et al., 1957) and in the pion decay chain $\pi \to \mu\nu, \mu \to e\nu\nu$ (Garwin
et al., 1957; Friedman and Telegdi, 1957). The theoretical and experimental
activity stimulated by Lee and Yang's conjecture led to a generalisation of Fermi's
vector theory, independently proposed by Feynman and Gell-Mann (1958) and
by Sudarshan and Marshak (1958). In their scheme, weak processes are described
by an effective Lagrangian density in which a universal charged weak current J^λ,
with a Lorentz structure containing both vector and axial vector parts $(V - A)$,
is coupled to itself at a single space-time point

$$\mathcal{L}_{\text{eff}} = -\frac{4G}{\sqrt{2}} J_\lambda J^{\lambda\dagger} + \text{h.c.} .\tag{2.1}$$

G is the Fermi coupling constant, determined from μ decay (see e.g. Sirlin, 1984):

$$G = (1.16632 \pm 0.00002) \times 10^{-5} [\text{GeV}^{-2}] .\tag{2.2}$$

The current J^λ consists of a leptonic and a hadronic part

$$J^\lambda = J^\lambda(l) + J^\lambda(h) .\tag{2.3}$$

The leptonic part $J^\lambda(l)$ is a straightforward generalisation from β decay, including the known leptons

$$J^\lambda(l) = \bar{e}\gamma^\lambda \frac{1-\gamma_5}{2}\,\nu_e + \bar{\mu}\gamma^\lambda \frac{1-\gamma_5}{2}\,\nu_\mu + \bar{\tau}\gamma^\lambda \frac{1-\gamma_5}{2}\,\nu_\tau \,. \qquad (2.4)$$

In this and the following expressions, the particle symbols denote the Dirac spinors for the different fermions involved in the interaction.

The $V - A$ structure of the charged weak current in the Feynman and Gell-Mann scheme can be directly read from the coupling of the spinors involved. The purely left-handed nature of the neutrino is introduced by the chirality operator $(1-\gamma_5)/2$, which projects the left-handed state. Working in the ultra-relativistic limit the chirality operator is identical with the helicity operator, which will be used extensively lateron.

In the light of present knowledge, the hadronic charged weak current is expressed in terms of quark fields

$$J^\lambda(h) = \bar{u}\gamma^\lambda \frac{1-\gamma_5}{2}\,d_C + \dots \,, \qquad (2.5)$$

where d_C describes the Cabibbo-rotated quark field to account for the relative coupling strengths in strangeness-changing and -conserving reactions (Cabibbo, 1963):

$$d_C = d\cos\theta_C + s\sin\theta_C \,. \qquad (2.6)$$

θ_C is called the Cabibbo angle and is experimentally determined as $\theta_C \sim 13.2°$ (see Particle Data Group, 1982).

The Feynman and Gell-Mann scheme, supplemented by Cabibbo's hypothesis, provided an excellent phenomenological description of the measured charged weak interactions within the framework of first-order perturbation theory. But it suffered from the same fundamental difficulty which was already encountered with the Fermi theory: the current-current interaction (2.1) inevitably violates unitarity (conservation of probability) at energies of order $1/\sqrt{G}$.

The Lagrangian (2.1) with currents such as (2.4) can be viewed as the low energy limit of the weak interaction mediated by the exchange of new field quanta, the intermediate vector bosons W. These bosons, in contrast to the quantum of the electromagnetic interaction, the photon, have to be charged and massive (due to the short range character of the weak interaction). Unitarity can be saved in lowest order by introducing such heavy intermediate vector bosons, named $W^\pm$. But once these $W^\pm$ bosons were postulated one could think up reactions such as $\nu\bar{\nu} \to W^+W^-$ which again violate unitarity if the produced W's are longitudinally polarised. Furthermore higher order terms involving W's lead to infinities in the cross section: The theory is non-renormalisable. In quantum electrodynamics (QED) one also encounters divergent integrals from higher order diagrams. There, however, the divergences can be removed order by order which is ultimately connected with the concept of local gauge invariance. The requirement that the equation of motion for a particle with wave function ψ should be unchanged after a local $U(1)$ phase transformation

5

$$\psi'(\vec{x}, t) = e^{ieQ\chi(\vec{x},t)}\psi(\vec{x}, t) , \qquad (2.7)$$

leads to the covariant derivative in the equation of motion

$$D^\lambda = \partial^\lambda + ieQA^\lambda , \qquad (2.8)$$

where Q, the "generator" of the Abelian group $U(1)$, is the electric charge of the field ψ in units of the modulus of the electron charge e. D_μ replaces ∂_μ in the wave equations for the free particle to yield the corresponding equation for an interacting particle. Equation (2.8) gives rise to the familiar minimal coupling between the electromagnetic current $\bar{\psi}\gamma_\mu\psi$ and the gauge field A_μ describing the massless photon. When ψ is rotated into ψ' according to (2.7) the field A^λ transforms to $A^\lambda - \partial^\lambda\chi(\vec{x}, t)$ which leaves the physics (the electromagnetic fields) unchanged for arbitrary functions $\chi(\vec{x}, t)$.

Historically, when only charged current weak interactions were known, it seemed natural to investigate the class of non-Abelian gauge theories based on the weak isospin group $SU(2)$. Yang and Mills (1954) proposed such a theory incorporating local $SU(2)$ gauge transformations which consequently entailed a triplet of massless vector quanta, since three generators exist in $SU(2)$. As with ordinary $SU(2)$ of hadronic isospin, the members of a weak isomultiplet differ by one unit of charge. Therefore the gauge quanta mediating the transition between these states will carry the charges ± 1. They could be identified with the hypothetical $W^\pm$ bosons exchanged in charged current reactions. The third gauge boson would be electrically neutral. In an early attempt Schwinger (1957) tried to identify this boson with the photon, thereby unifying electromagnetic and weak interactions. To account for the short range of the weak interaction the charged $W^\pm$ bosons had to acquire mass via some mechanism, while their neutral partner W^0 $(= \gamma)$ had to remain massless. Invoking additional scalar and pseudoscalar fields with appropriate couplings, massive $W^\pm$'s and a massless γ could be arranged for. But there was no constraint on the mass of the $W^\pm$, nor did neutral weak currents exist in Schwinger's model. An alternative model with heavy W^0's was proposed by Bludman (1958). As in Schwinger's model, W masses were put in ad hoc, and neutral currents with purely left-handed $(V - A)$ couplings were predicted, in contrast to the later experimental findings. However, the attractive feature of unifying the electromagnetic and weak interactions was lost. Moreover, both schemes, due to the introduction of mass terms in the Lagrangian, manifestly violated gauge invariance and the corresponding theories were thus, again, unrenormalisable.

2.2 The Model

Two more ingredients were necessary to arrive at the standard theory. The first one originated from Glashow (1961) – similar ideas were advanced by Salam and Ward (1964) – who suggested enlarging the Schwinger–Bludman $SU(2)$ schemes by an additional $U(1)$ gauge group, which leads to an $SU(2) \times U(1)$ group structure. The new Abelian $U(1)$ group is associated with "weak hypercharge" just as $SU(2)$ was associated with "weak isospin". Glashow indeed proposed that the Gell-Mann and Nishijima relation also hold for the weak charges:

$$Q = (t_3 + y/2) \ . \tag{2.9}$$

Here Q is the electric charge of the weakly interacting particle, t_3 its third component of weak isospin $\vec{t}$, and y its weak hypercharge. Choosing $SU(2) \times U(1)$ as the gauge group leads to the covariant derivative (compare to the QED analogue (2.8)):

$$D^\lambda = \partial^\lambda + \mathrm{i}\frac{g}{2}\vec{\tau} \cdot \vec{W}^\lambda + \mathrm{i}\frac{g'}{2} y B^\lambda \ . \tag{2.10}$$

The generators $\vec{\tau}$ are the familiar Pauli matrices. $\vec{W}^\lambda, g$ and B^λ, g' are the gauge fields and coupling constants corresponding to $SU(2)$ and $U(1)$, respectively (∂^λ and the term proportional to B^λ are understood to be multiplied by the 2×2 unit matrix). The theory contains both the unification and the weak neutral current aspects: The physical states for the photon A^λ and the neutral weak vector boson Z^λ are obtained by mixing the fields $W^{3\lambda}$ (the third component of $\vec{W}^\lambda$, also named W^0 above) and B^λ

$$A^\lambda = \sin\theta_{\mathrm{W}}\ W^{3\lambda} + \cos\theta_{\mathrm{W}}\ B^\lambda \ ,$$

$$Z^\lambda = \cos\theta_{\mathrm{W}}\ W^{3\lambda} - \sin\theta_{\mathrm{W}}\ B^\lambda, \tag{2.11}$$

where θ_{W} is an arbitrary mixing angle, now called the Weinberg angle.

The second essential ingredient was the concept of spontaneous symmetry breaking which provided a way to give mass to the intermediate bosons without destroying gauge invariance (the Higgs mechanism: Higgs, 1964; Englert and Brout, 1964; Guralnik et al., 1964; Kibble, 1967). The basic idea is to introduce a new (complex) scalar field ϕ, the Higgs field, which couples to the gauge fields in a way prescribed by the covariant derivative (see (2.8, 10)). Choosing a non-zero vacuum expectation value for ϕ (thereby "spontaneously breaking" the symmetry, better: hiding the gauge invariance) leads to the equations of motion for massive gauge quanta.

The crucial importance of symmetry breaking for a theory of the weak interactions was independently recognized by Weinberg (1967) and Salam (1968). Taking up Glashow's $SU(2) \times U(1)$ gauge model with its mixing relation (2.11) they proposed a minimal symmetry breaking scheme where a scalar weak isodoublet field ϕ

$$\phi = \begin{pmatrix} \phi_1 + \mathrm{i}\phi_2 \\ \phi_3 + \mathrm{i}\phi_4 \end{pmatrix} = \begin{pmatrix} \phi^+ \\ \phi^0 \end{pmatrix} \tag{2.12}$$

assumes a non-vanishing vacuum expectation value

$$\langle\phi\rangle_0 = \begin{pmatrix} 0 \\ v/\sqrt{2} \end{pmatrix} \tag{2.13}$$

for the lower, electrically neutral, component of the weak isospinor. The other three degrees of freedom of the complex scalar field are "eaten up" as to give mass to $W^\pm$ and Z^0, the photon remains massless. The Higgs particle is also instrumental in giving mass to the fermions. This is achieved by adjusting the fermion couplings to the Higgs proportional to the fermion masses.

A definite prediction of the scheme is thus the existence of a neutral scalar field, the Higgs boson, with well-defined couplings to the elementary fermions and to the weak gauge bosons. The masses of the weak bosons are given in terms of the vacuum expectation value v and the gauge couplings g and g':

$$M_W = \frac{g}{2}\, v,$$
$$M_Z = \frac{\sqrt{g^2 + g'^2}}{2}\, v\ . \tag{2.14}$$

With the mixing angle θ_W defined as

$$\tan\theta_W = \frac{g'}{g}\ , \tag{2.15}$$

the boson masses M_W, M_Z are consequently related through

$$\cos\theta_W = \frac{M_W}{M_Z}\ . \tag{2.16}$$

By insertion of (2.9, 11, 15) into (2.10) one obtains for the covariant derivative

$$D^\lambda = \partial^\lambda + \frac{i\,g}{\sqrt{2}}\left(\tau^+ W^{-\lambda} + \tau^- W^{+\lambda}\right)$$
$$+\, i\,g\sin\theta_W Q A^\lambda + \frac{i\,g}{\cos\theta_W}(t_3 - Q\sin^2\theta_W)\, Z^\lambda\ . \tag{2.17}$$

The second term in (2.17) describes the weak charged current interaction with the isospin raising and lowering operators $\tau^\pm = (\tau_1 \pm i\tau_2)/2$, while the last two terms represent the electromagnetic and weak neutral current interactions. The unification is evident by comparing (2.17) with the covariant derivative for the purely electromagnetic interaction (2.8). This leads to the following relation between the electromagnetic (e) and $SU(2)$ (g) coupling constants:

$$e = g\sin\theta_W\ . \tag{2.18}$$

With the "unification" relation (2.18) the coupling constants $C(t_3, Q)$ of the fermions to the Z^0 field in (2.17) become

$$C(t_3, Q) = \frac{ie}{\sin\theta_W \cos\theta_W} c(t_3, Q) , \qquad (2.19)$$

$$c(t_3, Q) = t_3 - Q\sin^2\theta_W . \qquad (2.20)$$

The coefficients $c(t_3, Q)$ in (2.20) are also called chiral coupling constants. If parity conservation is violated in the neutral current interaction these coupling constants will assume different values for the different chirality (left-handed and right-handed) states. Right-handed neutrinos, e.g., can be forbidden by construction if such neutrinos are placed in singlets of weak isospin ($t = 0$). The chiral coupling vanishes according to (2.20), leading to maximal parity violation in neutrino-induced neutral current interactions. Massive particles can exist in both chirality states so that parity violation occurs at a reduced level, depending on the actual size of the chiral couplings in (2.20). Charged current transitions, on the other hand, proceed only within the weak isospin doublets and are thus limited to left-handed particles. The coupling strength is given by the $SU(2)$ parameter g. All coupling constants for the charged and neutral current interactions are assumed to be the same for the different generations ("Universality").

Once the gauge group is chosen, the various particles participating in the interaction will have to be assigned to specific weak isospin multiplets in order to calculate their couplings to the weak currents. In the Glashow-Salam-Weinberg (GSW) theory the leptons are grouped into generations or families with two members each: A charged lepton l and its associated neutrino ν_l, where l stands for e, μ, τ, and possible further "sequential" leptons. The members of each family in turn, depending on their helicity state with respect to a given quantisation axis, are assigned to a left-handed weak isodoublet L or a right-handed weak isosinglet R:

$$
\begin{aligned}
L(l) &= \frac{1}{2}(1 - \gamma_5)\begin{pmatrix} \nu_l \\ l \end{pmatrix} & Q &= 0, & t_3 &= +1/2, \\
& & Q &= -1, & t_3 &= -1/2, \\
R(l) &= \frac{1}{2}(1 + \gamma_5)\, l & Q &= -1, & t_3 &= 0, \\
R(\nu_l) &= 0 & Q &= 0, & t_3 &= 0.
\end{aligned}
\qquad (2.21)
$$

Both the charge and the third component of the weak isospin are given in (2.21) so that the weak hypercharge for each particle state can be obtained from (2.9). Note that the right-handed neutrino spinor $R(\nu_l)$ vanishes.

The quarks are grouped into weak isomultiplets in an entirely analogous way. Since hadronic charged current transitions proceed only among the left-handed quarks, the left-handed members of the first generation of quarks, denoted by q_1 and containing the u and the d quark, are put into a weak isospin doublet:

$$
L(q_1) = \frac{1}{2}(1 - \gamma_5)\begin{pmatrix} u \\ d_C \end{pmatrix} \qquad
\begin{aligned}
Q &= +2/3, & t_3 &= +1/2, \\
Q &= -1/3, & t_3 &= -1/2,
\end{aligned}
\qquad (2.22)
$$

where d_C is the Cabibbo-rotated quark field (2.6). However, there was a serious flaw of the model by the time it was formulated: Neutral currents involving the quark field d_C are predicted to produce strangeness-changing transitions, which can be verified by forming the current

$$\bar{d}_C\, d_C = \left(\bar{d}\cos\theta_C + \bar{s}\sin\theta_C\right)(d\cos\theta_C + s\sin\theta_C)$$

$$= \bar{d}d\,\cos^2\theta_C + \bar{s}s\sin^2\theta_C + \left(\bar{d}s + \bar{s}d\right)\cos\theta_C\sin\theta_C\,. \tag{2.23}$$

According to (2.23) a strangeness-changing neutral current $(\bar{d}s+\bar{s}d)$, comparable in strength to the non-changing currents $\bar{d}d$ and $\bar{s}s$, is predicted. This is in strong contradiction to the exceedingly small branching ratio for $K^0_L \to \mu^+\mu^-$.

A revolutionary solution to this problem was proposed by Glashow, Iliopoulos and Maiani (1970), referred to as the GIM mechanism: GIM postulated the existence of a second doublet, consisting of a then undiscovered quark of charge $Q = +2/3$, the charmed quark, and a combination s_C of d and s quarks, orthogonal to d_C

$$L(q_2) = \frac{1}{2}(1 - \gamma_5)\begin{pmatrix} c \\ s_C \end{pmatrix} \qquad \begin{array}{ll} Q = +2/3\,, & t_3 = +1/2\,, \\ Q = -1/3\,, & t_3 = -1/2\,, \end{array} \tag{2.24}$$

with

$$s_C = -d\sin\theta_C + s\cos\theta_C\,. \tag{2.25}$$

Summing up both parts d_C and s_C and forming the neutral current from the available $q\bar{q}$ states $(\bar{d}_C\, d_C + \bar{s}_C\, s_C)$ leads to a cancellation of the forbidden flavour-changing $\bar{d}\,s$ and $\bar{s}\,d$ terms: The Z^0 thus couples only to $\bar{d}\,d$ and $\bar{s}\,s$, i.e. "diagonal in flavour". In addition to the prediction of the charmed quark, GIM could also limit the charm mass between 1 and 3 GeV by comparing the experimental $K^0_L \to \mu^+\mu^-$ decay rate to the prediction for the decay via two W's, which is the lowest order allowed process. In 1974 the charm or c quark, with properties predicted by GIM, was discovered (Aubert et al., 1974; Augustin et al., 1974).

The discovery of the τ lepton (Perl et al., 1975) and the bottom or b quark of charge $Q = -1/3$ (Herb et al., 1977) suggests the existence of a third generation of quarks and leptons. Attributing the b quark to the $t_3 = -1/2$ state of a weak left-handed doublet thus predicts the existence of another quark t or top quark with charge $Q = +2/3$.

Concerning the weak decays of the quarks of the second and third generation, the Cabibbo rotation matrix (see (2.6, 25)) can be generalised as suggested by Kobayashi and Maskawa (1973)

$$\begin{pmatrix} d' \\ s' \\ b' \end{pmatrix} = \begin{pmatrix} c_1 & s_1 c_3 & s_1 s_3 \\ -s_1 c_2 & c_1 c_2 c_3 - s_2 s_3 e^{i\delta} & c_1 c_2 c_3 + s_2 c_3 e^{i\delta} \\ s_1 s_2 & -c_1 s_2 c_3 - c_2 s_3 e^{i\delta} & -c_1 s_2 s_3 + c_2 c_3 e^{i\delta} \end{pmatrix} \begin{pmatrix} d \\ s \\ b \end{pmatrix}. \tag{2.26}$$

In the Cabibbo-Kobayashi-Maskawa (CKM) matrix the variables c_i, s_i denote $\cos\theta_i$, $\sin\theta_i$, with θ_1 equal to the Cabibbo angle θ_C. The angles $\theta_{1,2}$ turn out to be small experimentally (see e.g. Barbiellini and Santoni (1986) for a recent review). As for the leptons, the right-handed quark fields are considered to be weak isospin singlets:

$$R(q^i) = \frac{1}{2}(1+\gamma_5)\,q^i, \qquad i = \begin{cases} u,c,t \;:\; Q=+2/3\,, & t_3=0\,, \\ d,s,b \;:\; Q=-1/3\,, & t_3=0\,. \end{cases} \tag{2.27}$$

Using the above expressions for the left-handed and right-handed fields the weak neutral transition current for any fermion f (lepton or quark) can be obtained from the covariant derivative (2.17):

$$\langle f|J_{\mathrm{NC}}^\lambda|f\rangle = \frac{e}{\sin\theta_W\cos\theta_W}J_{\mathrm{NC}}^\lambda(f)\,, \tag{2.28}$$

with

$$J_{\mathrm{NC}}^\lambda(f) = \bar{f}\gamma^\lambda\left(c_{\mathrm{L}}\frac{1-\gamma_5}{2}+c_{\mathrm{R}}\frac{1+\gamma_5}{2}\right)f\,, \tag{2.29}$$

where c_{L} and c_{R} are the left- and right-handed chiral couplings as given by (2.20). Inserting (2.20) into (2.29) yields the familiar expression for the weak neutral transition current:

$$J_{\mathrm{NC}}^\lambda(f) = J^{3\lambda}(f) - \sin^2\theta_W\,J_{\mathrm{em}}^\lambda(f)\,, \tag{2.30}$$

where

$$J^{3\lambda}(f) = \bar{f}\gamma^\lambda t_3\frac{1}{2}(1-\gamma_5)f \tag{2.31}$$

is the third component of the weak isospin current and

$$eJ_{\mathrm{em}}^\lambda(f) = eQ_f\bar{f}\gamma^\lambda f \tag{2.32}$$

is the electromagnetic current. Equation (2.30) shows that the weak neutral current does not have a pure $V-A$ structure due to the additional vector piece from the electromagnetic current.

Transition amplitudes in lowest order are formed by sandwiching the gauge boson propagator between two transition currents. For the weak neutral current one thus arrives at the following expression for the transition amplitude A_{GSW}, taking the example of e^+e^- annihilation state into an arbitrary $f\bar{f}$ final state:

$$A_{\mathrm{GSW}} = \frac{e^2}{\sin^2\theta_W\cos^2\theta_W}J_{\mathrm{NC}}^\lambda(e)\frac{1}{s-M_Z^2+iM_Z\Gamma_Z}J_{\mathrm{NC}\,\lambda}(f)\,. \tag{2.33}$$

Table 2.1. Coupling constants of leptons and quarks to the neutral current

	$\nu_e,\ \nu_\mu,\ \nu_\tau$	$e,\ \mu,\ \tau$	$u,\ c,\ t$	$d,\ s,\ b$
Q_f	0	-1	$+2/3$	$-1/3$
$c_{\rm L}$	$\frac{1}{2}$	$-\frac{1}{2} + \sin^2\theta_{\rm W}$	$\frac{1}{2} - \frac{2}{3}\sin^2\theta_{\rm W}$	$-\frac{1}{2} + \frac{1}{3}\sin^2\theta_{\rm W}$
$c_{\rm R}$	0	$\sin^2\theta_{\rm W}$	$-\frac{2}{3}\sin^2\theta_{\rm W}$	$\frac{1}{3}\sin^2\theta_{\rm W}$
v	1	$-1 + 4\sin^2\theta_{\rm W}$	$1 - \frac{8}{3}\sin^2\theta_{\rm W}$	$-1 + \frac{4}{3}\sin^2\theta_{\rm W}$
a	1	-1	1	-1

For comparison, the corresponding QED amplitude is given by

$$A_{\rm QED} = -e^2 Q_f J_{\rm em}^\lambda(e)\frac{1}{s} J_{{\rm em}\,\lambda}(f) \, , \tag{2.34}$$

where the negative sign results from setting $Q_e = -1$.

In the local (low energy) limit the effective Lagrangian for the neutral current reads, in analogy to the charged currents in (2.1),

$$\mathcal{L}_{\rm eff}^{\rm NC} = -4G\sqrt{2}\, J_{{\rm NC}\,\lambda} \cdot J_{\rm NC}^{\lambda\dagger} + {\rm h.c.} \, , \tag{2.35}$$

where the constants e and $\theta_{\rm W}$ are related to the Fermi constant by means of (2.16, 18). Depending on certain assumptions in the Higgs sector of the theory, an additional parameter ϱ (see below) appears as a factor in (2.35), which is equal to unity in the simplest ("minimal") model.

The $SU(2) \times U(1)$ gauge theory of Glashow, Salam and Weinberg outlined above, attributing left-handed fermions to weak isodoublets and right-handed fermions to isosinglets, furthermore breaking the symmetry in a minimal way by means of a Higgs doublet, will be referred to in the following as the "standard theory" of electroweak interactions, or GSW theory. Often the group structure is written as $SU(2)_L \times U(1)$ in order to remind us that only left-handed weak isodoublets are allowed. In the standard theory the weak couplings (or charges) for leptons and quarks with respect to the neutral current are uniquely determined, as given in Table 2.1.

Rearranging terms of the transition current (2.29) into vector (γ^λ) and axial vector ($\gamma^\lambda\gamma_5$) parts leads to the definition of vector and axial vector charges v_i and a_i, which are also given in the Table 2.1. The constants v_i and a_i are related to $c_{\rm L}(i)$ and $c_{\rm R}(i)$ via

$$v_i = 2\left(c_{\rm L}(i) + c_{\rm R}(i)\right) ,$$
$$a_i = 2\left(c_{\rm L}(i) - c_{\rm R}(i)\right) . \tag{2.36}$$

To complete this section it should be mentioned without further details that the GSW theory could be proven to be renormalisable (t'Hooft, 1971 a,b; Lee, 1972a; Lee and Zinn-Justin, 1972a—c). Feynman rules have been developed (t'Hooft and Veltman, 1972a,b) which provide the basis for practical perturbation calculations, in particular for radiative and higher order weak corrections (see Chaps. 3 and 7).

2.3 Predictions

The GSW theory, supplemented by the GIM or CKM prescription of quark mixing, makes a number of firm predictions which were by no means substantiated experimentally by the time of formulation:

— Existence of neutral current reactions in lowest order by virtue of exchange of a neutral weak boson. The process should be comparable in strength to the charged current reactions. Only one additional free parameter, the Weinberg angle θ_W, is introduced.
— Existence of a new quark called charm with charge +2/3 and mass between 1 and 3 GeV/c^2.
— Existence of a triplet of heavy gauge bosons, $W^\pm$ and Z^0, with masses depending essentially on the value of the Weinberg angle $\sin^2 \theta_W$, and of a heavy scalar particle, the Higgs boson.
— Universality of chiral couplings of the weak neutral current with respect to the existing generations of leptons and quarks. Given the fine structure constant α and Fermi's constant G, all couplings are determined, in lowest order, by the Weinberg angle $\sin^2 \theta_W$, and an overall scale parameter ϱ which is equal to 1 for the "minimal" Higgs scheme (see below).

The list of discoveries following the predictions was in fact impressive:

— Discovery of weak neutral currents by observation of the reaction $\nu_\mu\, e \rightarrow \nu_\mu\, e$ in a heavy liquid bubble chamber (Hasert et al., 1973).
— Discovery of charm in $p\,Be$ reactions (Aubert et al., 1974) and in $e^+ e^-$ annihilation (Augustin et al., 1974), the famous "November Revolution".
— Discovery of the heavy gauge bosons $W^\pm$ and Z^0 in $\bar{p}p$ reactions (Arnison et al., 1983a; Banner et al., 1983; Arnison et al., 1983b; Bagnaia et al., 1983).

What remained to be done on the experimental side was to

— measure the free parameter of the theory, $\sin^2 \theta_W$,
— test universality,
— verify the Higgs scheme: Measure the ϱ parameter, and/or discover the Higgs scalar.

Given the value for $\sin^2 \theta_W$ all couplings are fixed up to a common factor ϱ. The expected couplings for $\sin^2 \theta_W = 0.23$ and $\varrho = 1$ (which is strongly suggested by

Table 2.2. Coupling constants of the elementary fermions to the weak neutral current for $\sin^2 \theta_W = 0.23$

	ν_e, ν_μ, ν_τ	e, μ, τ	u, c, t	d, s, b
Q_f	0	-1	$+2/3$	$-1/3$
c_L	0.5	-0.27	0.35	-0.42
c_R	0	0.23	-0.15	0.08
v	1	-0.08	0.40	-0.69
a	1	-1	1	-1

a combined analysis of all available neutral current data, see below) are given in Table 2.2, where universality is explicitely assumed. As will be demonstrated in this review almost all couplings in the table have been measured, in agreement with the predictions, and measurements of $\sin^2 \theta_W$ and ϱ from the various reactions are consistent.

Somewhat more uncertain are the predictions for the Higgs scalar. The standard theory assumes the simplest Higgs structure, i.e. a weak isodoublet of Higgs scalars, thus often referred to as the minimal model (see (2.12)). In principle, representations of higher dimensionality for the Higgs scalars are possible. Their effect on the mass spectrum of the intermediate vector bosons and the strength of the neutral current interaction is characterized by the parameter ϱ, multiplying the Lagrangian density (2.35)

$$\varrho = \frac{M_W^2}{M_Z^2 \cos^2 \theta_W} \, . \tag{2.37}$$

With the Higgs scalar doublets of the standard theory ϱ is equal to unity (see (2.16)). More generally, ϱ depends on the weak isospin of the Higgs scalar fields in the following way (Lee, 1972b):

$$\varrho = \frac{\sum \left[t(t+1) - t_3^2\right] \langle \phi_{t,t_3} \rangle_0^2}{2 \sum t_3^2 \langle \phi_{t,t_3} \rangle_0^2} \, , \tag{2.38}$$

where the sum is taken over the different Higgs fields of weak isospin t, t_3, with vacuum expectation values $\langle \phi_{t,t_3} \rangle_0$. Since ϱ is a scale factor of the neutral weak currents with respect to the charged weak currents, it can be determined by experiment.

The most direct way of testing the standard theory certainly is to search for the massive bosons themselves. For the estimate of masses one may compare the effective coupling constant in (2.1) with the coefficients multiplying the charged current in (2.17) which yields the lowest order relation between G and g:

$$\frac{4G}{\sqrt{2}} = \frac{g^2}{2M_W^2} , \tag{2.39}$$

where the W boson propagator term, similar to (2.33, 35), has been taken in the low energy limit. Inserting the unification condition (2.18) and the general relation (2.37) between M_W and M_Z one obtains, using the fine structure constant $\alpha = e^2/4\pi$

$$\frac{G}{\sqrt{2}} = \frac{\pi\alpha}{2\sin^2\theta_W M_W^2} ,$$

$$\varrho\frac{G}{\sqrt{2}} = \frac{\pi\alpha}{2\sin^2\theta_W \cos^2\theta_W M_Z^2} . \tag{2.40}$$

In the standard theory the masses of the heavy bosons are thus predicted from a measurement of the Weinberg angle. For a precise estimate of the boson masses, however, one has to take into account radiative corrections which can, taking the case of the Z^0 from (2.40), be written as

$$M_Z^2 = \frac{\pi\alpha}{\varrho G\sqrt{2}\sin^2\theta_W \cos^2\theta_W(1 - \delta r)} . \tag{2.41}$$

The radiative correction δr has been calculated, e.g., by Marciano and Sirlin (1984) and is found to be about 0.07, depending on the masses of the Higgs boson, the t quark and the values for the Weinberg angle. With this numerical correction the boson masses are given by

$$M_W = \frac{38.68\,[\text{GeV}/c^2]}{\sin\theta_W} ,$$

$$M_Z = \frac{77.37\,[\text{GeV}/c^2]}{\sqrt{\varrho}\,\sin 2\theta_W} . \tag{2.42}$$

Taking the experimental value $\sin^2\theta_W = 0.23$ (see the following chapters), the standard theory ($\varrho = 1$) makes the definite predictions

$$M_W = 80.7\,[\text{GeV}/c^2],$$

$$M_Z = 91.9\,[\text{GeV}/c^2] . \tag{2.43}$$

The widths Γ_Z and Γ_W of the weak bosons are also calculable. Using the matrix element (2.29) the partial width of the Z^0 decay to a fermion-antifermion pair is given by

$$\Gamma(Z^0 \to f\bar{f}) = \frac{G}{6\pi\sqrt{2}} M_Z^3 \frac{(v_f^2 + a_f^2)}{4} , \tag{2.44}$$

15

where v_f and a_f are the vector and axial vector coupling constants of the produced fermion (see (2.36) and Table 2.1). A similar calculation for the $W^\pm$ into a left-handed charged fermion-antifermion pair yields

$$\Gamma(W^\pm \to f\bar{f}') = \frac{G}{6\pi\sqrt{2}} M_W^3 \cdot \qquad (2.45)$$

With six quarks and six leptons the sum over all possible decay channels for the Z^0 and $W^\pm$ (including the colour factor of 3 for decays into quark-antiquark pairs) leads to the total widths of $\Gamma_Z \simeq 2.8\,\mathrm{GeV}/c^2$ and $\Gamma_W \simeq 2.7\,\mathrm{GeV}/c^2$. Both the discovery of $W^\pm$ and Z^0 and a measurement of their masses are evidently crucial experimental cornerstones of the standard theory. The knowledge of the widths furthermore yields information on the number of light fermions (including neutrinos) when the standard couplings (2.44) are assumed for each of them. In case the t quark is heavier than about 45 GeV, the width of the Z^0 is reduced by only about 0.3 GeV.

3. Electroweak Effects in e^+e^- Interactions

High energy e^+e^- interactions have proven to be extremely useful tools during the past few years for the investigation.of neutral current phenomena. There are various reasons why for the continued interest in e^+e^- interactions. In the e^+e^- annihilation into a single photon, fermion-antifermion pairs are produced in proportion to the square of their electric charge and their colour multiplicity, thus providing access to leptons and quarks (with mass less than the beam energy) from all generations. In particular, the production of leptons and quarks of the second and third generation makes e^+e^- annihilation a unique tool for testing the universal character of the standard theory. The simple and known initial state kinematics facilitates the study of more complex decays in the final state, e.g. heavy quarks and τ leptons. The latter, in fact, could only be isolated so far in e^+e^- reactions.

The neutral current is studied mainly through its interference with the electromagnetic current which still dominates at the presently accessible centre of mass energies, well below the threshold for Z^0 formation. High energy storage rings such as PEP or PETRA have provided good luminosity (still a factor of ~ 100 better than the $\bar{p}p$ collider) at timelike momentum transfers squared Q^2 up to ~ 2200 $(\text{GeV}/c)^2$, corresponding to $\sqrt{s} = 46.78$ GeV (top end of PETRA). Recently, TRISTAN has come into operation pushing the explorable range to $Q^2 \sim 3000$ $(\text{GeV}/c)^2$. The effects due to electroweak interference are of order Q^2/M_Z^2 and one thus expects clear departures from pure quantum electrodynamics at the 10% to 20% level.

3.1 Electroweak Phenomenology

The purpose of this chapter is to derive the necessary formulae describing e^+e^- reactions in the standard theory of electroweak interactions. In lowest order QED the e^+e^- annihilation into charged fermions f is mediated by the s-channel exchange of a time-like photon, as depicted in Fig. 3.1. For $f = e^-$ (Bhabha scattering) there is an additional graph describing the t-channel exchange of a spacelike photon. Bhabha scattering will be treated in Sect. 3.4.1. The lowest order diagram due to the weak neutral current in the GSW theory is shown in Fig. 3.2. For the evaluation of reaction cross sections the corresponding amplitudes A_{QED} and A_{GSW}, given in (2.33, 34), are added coherently. Inserting the

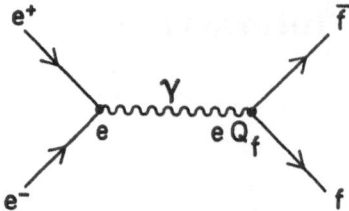

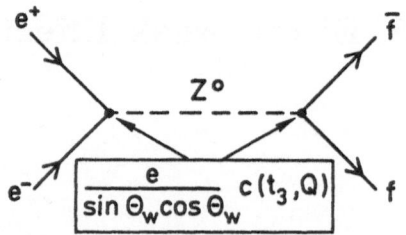

Fig. 3.1. Feynman diagram for the one-photon exchange in the s channel for the reactions $e^+e^- \to f\bar{f}$

Fig. 3.2. Feynman diagram for the exchange of a neutral weak boson (Z^0) in the s channel for the reactions $e^+e^- \to f\bar{f}$

transition currents (2.29, 32) yields the following expressions for the two transition amplitudes:

$$A_{\text{GSW}} = \left(\frac{e^2}{\sin^2\theta_{\text{W}}\cos^2\theta_{\text{W}}}\right)\left(\frac{1}{s - M_Z^2 + iM_Z\Gamma_Z}\right)$$
$$\cdot \left[\bar{\psi}_e\gamma^\lambda \left(c_{\text{L}}(e)\frac{1-\gamma_5}{2} + c_{\text{R}}(e)\frac{1+\gamma_5}{2}\right)\psi_e\right.$$
$$\left.\cdot \bar{\psi}_f\gamma_\lambda \left(c_{\text{L}}(f)\frac{1-\gamma_5}{2} + c_{\text{R}}(f)\frac{1+\gamma_5}{2}\right)\psi_f\right],$$

$$A_{\text{QED}} = \frac{e^2}{s}\cdot\left[\bar{\psi}_e\gamma^\lambda\psi_e \cdot \bar{\psi}_f\gamma_\lambda\psi_f\right],$$

(3.1)

where the particle spinors are denoted by ψ. Taking into account the usual flux factor for the incoming beams one can express the electroweak cross section σ_{EW}, up to a factor of order unity, as

$$\sigma_{\text{EW}}(e^+e^- \to f\bar{f}) \approx \frac{1}{s}|A_{\text{QED}} + A_{\text{GSW}}|^2.$$

(3.2)

For a rough estimate of the relative contributions from photon and Z^0 exchange one may consult the above expressions in (3.1) for the transition amplitudes. The amplitude A_{QED} is of order $\alpha \cdot s/s$, where s in the demominator comes from the photon propagator, and s in the numerator from the contraction of the Dirac spinors for the fermions involved (using the covariant normalisation $u^\dagger u = \sqrt{s}$ for the Dirac spinors). Similarly A_{GSW} is given, up to some trivial factors of order unity, by $\alpha \cdot s/M_Z^2$, where the factor s in the numerator again arises from the Dirac spinors. The denominator obviously is the low energy approximation ($s \ll M_Z^2$) for the Z^0 propagator term $1/(s - M_Z^2 + iM_Z\Gamma_Z)$.

As the ratio α/M_Z^2 is proportional to the Fermi coupling constant G (see (2.40)), the cross section (3.2) can be written as a sum of the pure QED, pure weak and the interference term, each with its characteristic energy dependence:

$$\sigma_{\text{EW}} = \sigma\left(\text{QED}, \sim \frac{\alpha^2}{s}\right) + \sigma(\text{interf.}, \sim \alpha G) + \sigma(\text{weak}, \sim G^2 s).$$

(3.3)

The QED cross section has the well-known $1/s$ dependence while the purely weak cross section rises linearly with s, neglecting propagator effects. As will be shown later, the purely weak contribution to the total cross section amounts to only a fraction of a percent of the QED cross section for leptonic final states at the energies accessible with PEP/PETRA. In contrast, the purely weak contribution to the total cross section for e^+e^- annihilation into quark pairs leading to hadrons is expected to yield a measurable effect.

The cleanest evidence for weak neutral currents in e^+e^- reactions, the forward-backward angular asymmetry, is caused mainly by the electroweak interference term at PEP/PETRA energies. This term, which rises like s relatively to QED, will provide the bulk of information about the weak neutral current. It is interesting to note that some of the information, such as the sign of the weak couplings, can only be obtained by studying electroweak interference (see below). Again the effects involving quarks are more pronounced but are also largely clouded in the observable multihadronic final state which is not necessarily specific for the parent quark-antiquark system.

The characteristic signature of the weak interaction is parity violation arising from a pseudoscalar observable, i.e. a quantity proportional to the product of a vector and an axial-vector piece in the reaction amplitude. Since the electron and positron beams in PEP/PETRA are longitudinally unpolarised, the parity-violating part of the electroweak interference amplitude can only be studied by measuring the polarisation of final state fermions. Particularly useful are reactions with τ leptons or heavy quarks in the final state, the weak decay of which could potentially be used to analyse their polarisation.

3.1.1 Differential Cross Section

Expressions for the various cross sections (3.2), which display the dependence on the weak neutral coupling constants as defined in Chap. 2, are most easily calculated from amplitudes describing states of definite helicity (see e.g. Sehgal, 1980). The helicity amplitudes F for the scattering of massless spin $1/2$ particles display a very simple dependence on the scattering angle θ between initial and final fermions in the centre of mass system. For initial state particles in the same helicity state, i.e. net helicity zero, one has $F \sim 1$ (isotropic scattering), while for opposite helicities $F \sim (1 \pm \cos\theta)/2$. The sign in front of the $\cos\theta$ term depends on the helicity configuration of the final state particles and will be clarified below.

The two-body cross section with incoming particles 1,2 and outgoing particles 3,4 is given by (see, e.g., Aitchison and Hey, 1982):

$$d\sigma = \frac{1}{4\sqrt{(P_1 \cdot P_2)^2 - m_1^2 m_2^2}} |A|^2 \, d\,\text{Lips}\,(s, P_3, P_4)\,, \qquad (3.4)$$

where P_i are the particle four vectors, m_i the particle masses, and $d\,\text{Lips}$ the

Lorentz-invariant two-body phase space element, and A is the transition matrix element. In the centre of mass system, neglecting all masses, (3.4) reduces to the following expression for the differential cross section with unpolarised beams:

$$\frac{d\sigma}{d\Omega} = \frac{1}{4} \cdot \frac{1}{16\pi^2 s} |A|^2 .$$

(3.5)

The factor $1/4$ arises from averaging over the particle spins S of the initial state particles, where each particle contributes a statistical weight factor $1/(2S+1)$. The amplitude A can be decomposed into an incoherent sum of amplitudes $\tilde{F}_{ij}$ for fixed helicities i, j in the initial and final state (Jacob and Wick, 1959):

$$|A|^2 = \sum |\tilde{F}_{ij}|^2 .$$

(3.6)

For reasons of economy in the discussion of the various neutral current processes, throughout this report a factor of $4\pi \alpha s$ will be extracted from the usual definition of the helicity amplitudes:

$$\tilde{F}_{ij} = 4\pi \alpha s F_{ij} .$$

(3.7)

This factor is motivated by the common coupling strength $e = \sqrt{4\pi\alpha}$ for the electromagnetic and weak neutral current (2.17, 18) and by the common factor s arising from the contraction of the fermion spinors (see above). The differential cross section (3.5) is thus given as

$$\frac{d\sigma}{d\Omega} = \frac{1}{4} \alpha^2 s \sum |F_{ij}|^2 .$$

(3.8)

Since helicity is conserved at each vertex in the diagrams depicted in Figs. 3.1, 2 due to the Lorentz structure of the electromagnetic (V) and the weak neutral current (V, A), the total helicity in the initial as well as the final state has to be ± 1. In the massless case helicity is equivalent to handedness, so that electrons and positrons only annihilate with opposite handedness (or helicity), i.e. $e_L^- e_R^+$ or $e_R^- e_L^+$. Choosing the helicity of, say, the fermion in the initial and final state thus fixes the helicity of the *anti*-fermion in the respective states. Consequently only four helicity amplitudes F_{ij} exist with i denoting the helicity (handedness) of the initial and j the handedness of the final fermion. These amplitudes are shown in Table 3.1, where θ is the scattering angle between initial and final fermion in the centre of mass system. The constants $c_i(e)$ and $c_j(f)$ are the chiral couplings (see (2.20) and Table 2.1). The angular dependence shown in Table 3.1 is a consequence of angular momentum conservation: $e_L^- e_R^+ \to f_L \bar{f}_R$ and $e_R^- e_L^+ \to f_R \bar{f}_L$ cannot occur at a scattering angle of $180°$. The two other processes $e_L^- e_R^+ \to f_R \bar{f}_L$ and $e_R^- e_L^+ \to f_L \bar{f}_R$ cannot occur at $0°$.

The functions $g(s)$ can be parametrized in two ways. In the first parametrisation, $g(s)$ depends on the Weinberg angle $\sin^2 \theta_W$ and M_Z (neglecting the small contribution of the Z^0 width far from the pole) and no connection is made with the charged current so far:

Table 3.1. Lowest order electroweak helicity amplitudes for s-channel processes $e^+e^- \rightarrow$ $f\bar{f}$ $(f = \mu^-, \tau^-, q)$

Helicity configuration	Amplitude F_{ij}
$e_L^- e_R^+ \rightarrow f_L \bar{f}_R$	$s^{-1}\varepsilon_{LL}(s)(1 + \cos\theta)/2$
$e_L^- e_R^+ \rightarrow f_R \bar{f}_L$	$s^{-1}\varepsilon_{LR}(s)(1 - \cos\theta)/2$
$e_R^- e_L^+ \rightarrow f_L \bar{f}_R$	$s^{-1}\varepsilon_{RL}(s)(1 - \cos\theta)/2$
$e_R^- e_L^+ \rightarrow f_R \bar{f}_L$	$s^{-1}\varepsilon_{RR}(s)(1 + \cos\theta)/2$
with $\varepsilon_{ij}(s) = Q_f - 8\,c_i(e)\,c_j(f)\,g(s)$	
$d\sigma/d\Omega = \frac{1}{4}\alpha^2 s \sum_{ij}|F_{ij}|^2$	

$$g(s) = \frac{1}{8\sin^2\theta_W \cos^2\theta_W}\left(\frac{s}{s - M_Z^2 + iM_Z\Gamma_Z}\right). \tag{3.9}$$

Alternatively, and closer to the spirit of the standard theory, relation (2.40) can be used, which "unifies" the electromagnetic and the weak interactions, to express the Weinberg angle in terms of the Fermi coupling G, the fine structure constant α, and the Z^0 mass

$$g(s) = \frac{\varrho\,G}{\sqrt{2}}\frac{M_Z^2}{4\pi\alpha}\left(\frac{s}{s - M_Z^2 + i\,M_Z\Gamma_Z}\right). \tag{3.10}$$

This expression for $g(s)$ was traditionally used in describing electroweak interference since, far from the Z^0 pole, all "unknown" parameters such as $\sin^2\theta_W$ and M_Z drop out:

$$g(s) = -\frac{G}{\sqrt{2}}\frac{s}{4\pi\alpha} \tag{3.11}$$

with

$$s \ll M_Z^2, \quad \varrho = 1.$$

One obtains for the differential cross section of a left-handed electron e_L^- $(+\,e_R^+)$ into a left-handed fermion $f_L(+\bar{f}_R)$:

$$\frac{d\sigma}{d\Omega}(e_R^+ e_L^- \rightarrow \bar{f}_R f_L) = \alpha^2 s|F_{LL}|^2 \tag{3.12}$$

with

$$F_{\mathrm{LL}} = \frac{Q_f - 8c_{\mathrm{L}}(e)c_{\mathrm{L}}(f)g(s)}{s}\left(\frac{1+\cos\theta}{2}\right).$$

Replacing the left-handed couplings $c_{\mathrm{L}}(e), c_{\mathrm{L}}(f)$ by $c_{\mathrm{R}}(e)$ and $c_{\mathrm{R}}(f)$ yields F_{RR}. The amplitude F_{LR} (and, correspondingly, F_{RL}) describing helicity flip transitions between initial and final state fermion and antifermion reads:

$$F_{\mathrm{LR}} = \frac{Q_f - 8c_{\mathrm{L}}(e)c_{\mathrm{R}}(f)g(s)}{s}\left(\frac{1-\cos\theta}{2}\right). \tag{3.13}$$

If the beams are unpolarised and final state polarisation is not observed, the differential cross section for $e^+e^- \to \bar{f}f$ is obtained by summing over the four squared amplitudes and averaging over initial helicity states (weight factor $1/4$):

$$\frac{4s}{\alpha^2}\frac{d\sigma}{d\Omega}(e^+e^- \to \bar{f}f)$$

$$= \frac{1}{4}\Big[|Q_f - 8c_{\mathrm{L}}(e)c_{\mathrm{L}}(f)g(s)|^2 + |Q_f - 8c_{\mathrm{R}}(e)c_{\mathrm{R}}(f)g(s)|^2\Big]$$

$$\cdot (1+\cos\theta)^2 \tag{3.14}$$

$$+ \frac{1}{4}\Big[|Q_f - 8c_{\mathrm{L}}(e)c_{\mathrm{R}}(f)g(s)|^2 + |Q_f - 8c_{\mathrm{R}}(e)c_{\mathrm{L}}(f)g(s)|^2\Big]$$

$$\cdot (1-\cos\theta)^2 .$$

By inspection of (3.14) it is clear that the cross section is invariant under parity transformation ($L \to R, R \to L$). Therefore any quantity derived from it such as the forward-backward angular asymmetry (see below) is intrinsically parity conserving. Using instead of the right- and left-handed couplings the vector and axial vector couplings $v_i, a_i (i = e, f)$ defined in (2.36), and sorting terms odd and even in $\cos\theta$, leads to the following expression:

$$\frac{4s}{\alpha^2}\frac{d\sigma}{d\Omega}(e^+e^- \to \bar{f}f) = C_{\mathrm{S}}(1+\cos^2\theta) + 2C_{\mathrm{A}}\cos\theta \tag{3.15}$$

with

$$C_{\mathrm{S}} = Q_f^2 - Q_f v_e v_f \operatorname{Re}\{g(s)\} + \frac{1}{4}(v_e^2 + a_e^2)(v_f^2 + a_f^2)|g(s)|^2 \tag{3.16}$$

and

$$C_{\mathrm{A}} = -Q_f a_e a_f \operatorname{Re}\{g(s)\} + v_e v_f a_e a_f |g(s)|^2 . \tag{3.17}$$

The coefficient C_{S}, describing the symmetric part of the angular distribution, is a sum of QED (proportional to Q_f^2), electroweak interference (proportional to $Q_f \cdot \operatorname{Re}\{g(s)\}$) and purely weak terms (proportional to $|g(s)|^2$), while only electroweak and purely weak terms contribute to the asymmetric part C_A. Radiative corrections to the lowest order QED cross section, however, will also produce an asymmetric part, as will be discussed in Sect. 3.2.

From (3.15) one can form an angular asymmetry, defined as

$$A(\theta) = \frac{\frac{d\sigma}{d\Omega}(\theta) - \frac{d\sigma}{d\Omega}(\pi - \theta)}{\frac{d\sigma}{d\Omega}(\theta) + \frac{d\sigma}{d\Omega}(\pi - \theta)}$$

$$= \frac{C_A}{C_S} \frac{2\cos\theta}{1 + \cos^2\theta} .$$

(3.18)

At the presently available centre of mass energies from PEP/PETRA, and even at TRISTAN, the purely weak terms in C_A and C_S are quite small ($\frac{1}{4}|g(s)|^2 \approx$ 0.004, 0.04 at $\sqrt{s} = 35$, 55 GeV) and Re$\{g(s)\}$ is ~ 0.1. Furthermore the electroweak interference term in C_S is multiplied by the vector coupling constant of the electron which is close to zero (see Table 2.1) and may thus be dropped. One then obtains the approximate formula for the angular asymmetry

$$A(\theta) \approx -\frac{a_e a_f}{Q_f} \text{Re}\{g(s)\} \frac{2\cos\theta}{1 + \cos^2\theta} .$$

(3.19)

By integrating the differential cross section (3.15) over the forward $(0, \pi/2)$ and backward $(\pi/2, \pi)$ hemispheres separately one derives an approximate expression for the forward-backward asymmetry A_{FB}:

$$A_{FB} = \frac{3}{4} \frac{C_A}{C_S} \approx -\frac{3}{4} \frac{a_e a_f}{Q_f} \text{Re}\{g(s)\} .$$

(3.20)

As only a limited range in $\cos\theta$ is usually accessible for particle detection in a real experiment , one measures the forward-backward asymmetry within the specific aceptance range of the experiment ($c = |\cos\theta_{max}|$) and then has to extrapolate to the full interval. In this case $A_{FB}(c)$ is reduced in modules with respect to A_{FB} and is related to (3.20) by

$$A_{FB}(c) = A_{FB} \frac{4c}{3 + c^2} .$$

(3.21)

The electroweak charge asymmetry A_{FB} is a function mainly of the axial vector coupling constants between the Z^0 and the fermions involved (which are ± 1 in the standard theory, see Table 2.1) and rises linearly in modulus with the square of the centre of mass energy for $M_Z \to \infty$ (see (3.11)). In fact, a finite mass of the Z^0 enhances the asymmetry in the PEP/PETRA range with respect to the linear rise (see (3.9, 10)). Furthermore, the asymmetry depends on the ratio a_f/Q_f of the fermion pair produced and is therefore larger for the fractionally charged quarks in comparison to the integrally charged leptons. For finite values of the Z^0 boson, the term $g(s)$ in (3.10) depends only weakly on the Weinberg angle $\sin^2\theta_W$

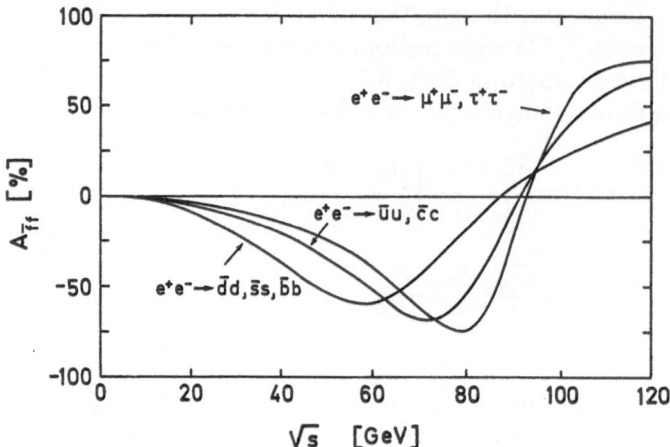

Fig. 3.3. Predictions from the standard theory for the charge asymmetry for fermion pair production in the e^+e^- annihilation as a function of the centre of mass energy. For the prediction the values $\sin^2\theta_W = 0.23$ and $M_Z = 92$ GeV have been used

through the Z^0 mass in the propagator, given as $M_Z = 77.4 \,[\text{GeV}/c^2]/\sin 2\theta_W$ (see (2.42)). Since the mass of the Z^0 is now known one may come back to relation (3.9) and try to extract the Weinberg angle from e^+e^- data, as will be done in Sect. 3.5.

The asymmetries for the charged fermions to be expected in the standard theory are illustrated in Fig. 3.3. Both for leptons l and quarks q the charge asymmetry in the entire range below the Z^0 is predicted to be negative, for leptons at the level of -10% at $\sqrt{s} \sim 35\,\text{GeV}$. For quarks with $Q_f = +\frac{2}{3}$ (u, c, t) one expects -15% and for quarks with $Q_f = -\frac{1}{3}$ (d, s, b) values as large as -30%.

However, with the more practical definition of the charge asymmetry using the angle between the incoming electron and the negative outgoing fermion (i.e. μ^-, τ^-, $\bar{u}$, $\bar{c}$, d, s, b, ...), the charge asymmetry for the up-type quarks changes sign. For the subsequent discussion a charge asymmetry using this above definition of the scattering angle will be called the signed charge asymmetry $\tilde{A}_{\text{FB}}(q)$. The average charge asymmetry for hadronic final states is then given by the individual signed quark charge asymmetries $\tilde{A}_{\text{FB}}(q)$, properly weighted by the production cross section, which is proportional to the square of the quark charge Q_q, i.e.

$$A_{\text{FB}}\left(\sum_q\right) = \frac{\sum_q Q_q^2 \, \tilde{A}_{\text{FB}}(q)}{\sum_q Q_q^2}. \tag{3.22}$$

If, in a given experiment, the separation of individual quark flavours from the hadronic final states is not possible, one consequently expects the charge asymmetries to partly cancel. Working out the combined asymmetry expected from (3.15) when u, d, c, s and b quarks are produced one obtains a largely reduced average quark charge asymmetry, sign-reversed with respect to the lepton charge asymmetry:

$$A_{\text{FB}}\left(\sum_q\right) = -\frac{3}{11} A_{\text{FB}}(\bar{l}l) \, . \tag{3.23}$$

The experimental problems involved in the determination of the quark charge (see, e.g., Maxwell and Teper, 1981) will further cloud the observable effect so that high statistics and good control of systematics are needed to perform such an inclusive measurement. On the other hand, attempts have been made (see below) to identify the quark flavour on an event-to-event basis ("flavour tagging") which cuts severely into the statistics, but displays the full effect.

3.1.2 Total Cross Section

Integrating (3.15), the terms odd in $\cos\theta$ drop out and one obtains the total cross section:

$$\sigma(e^+e^- \to \bar{f}f) = \frac{4\pi\alpha^2}{3s} C_{\text{S}} \, . \tag{3.24}$$

Besides the dominant QED part the total cross section $\sigma(e^+e^- \to \bar{f}f)$ receives contributions from electroweak interference proportional to $v_e v_f$ (very small, see Table 2.1) and purely weak terms in C_{S}. For leptonic final states purely weak effects on the total cross section are still unmeasurable even at the highest PETRA energies ($\sim +1.0\%$ at $\sqrt{s} = 46$ GeV). However, in the total hadronic cross section weak effects of about $+6\%$ are expected to show up at PETRA's top energy, due to the sizeable values for v_q^2 (see Table 2.1). If one assumes the validity of the quark parton model, which is well supported by the e^+e^- data, the total cross section is given by an incoherent sum over the contributions from the individual quark flavours

$$\sigma(e^+e^- \to \text{hadrons}) = \frac{4\pi\alpha^2}{3s} \, 3 \left[\sum_q Q_q^2 - v_e \, \text{Re}\{g(s)\} \sum_q Q_q v_q \right.$$
$$\left. + \frac{1}{4}(v_e^2 + a_e^2)|g(s)|^2 \sum_q (v_q^2 + a_q^2) \right] , \tag{3.25}$$

where kinematical terms due to the finite mass of the quarks have been neglected. The factor 3 in (3.25) accounts for the three coloured varieties of each quark. Additional contributions to $\sigma(e^+e^- \to \text{hadrons})$ are expected from higher order quantum chromodynamics (QCD) which are proportional to powers of the strong coupling constant α_s. As will be shown in Sect. 3.6.1, the analysis of weak contributions to the total cross section requires knowledge of the strong coupling constant.

The expected total cross sections for $e^+e^- \to \mu^+\mu^-$, $\tau^+\tau^-$ and $e^+e^- \to$ hadrons (no QCD corrections) are shown in Fig. 3.4, normalised to the lowest order QED cross section for $e^+e^- \to \mu^+\mu^-$, for various values of the Weinberg angle. As indicated above, the contributions from the weak propagator are very

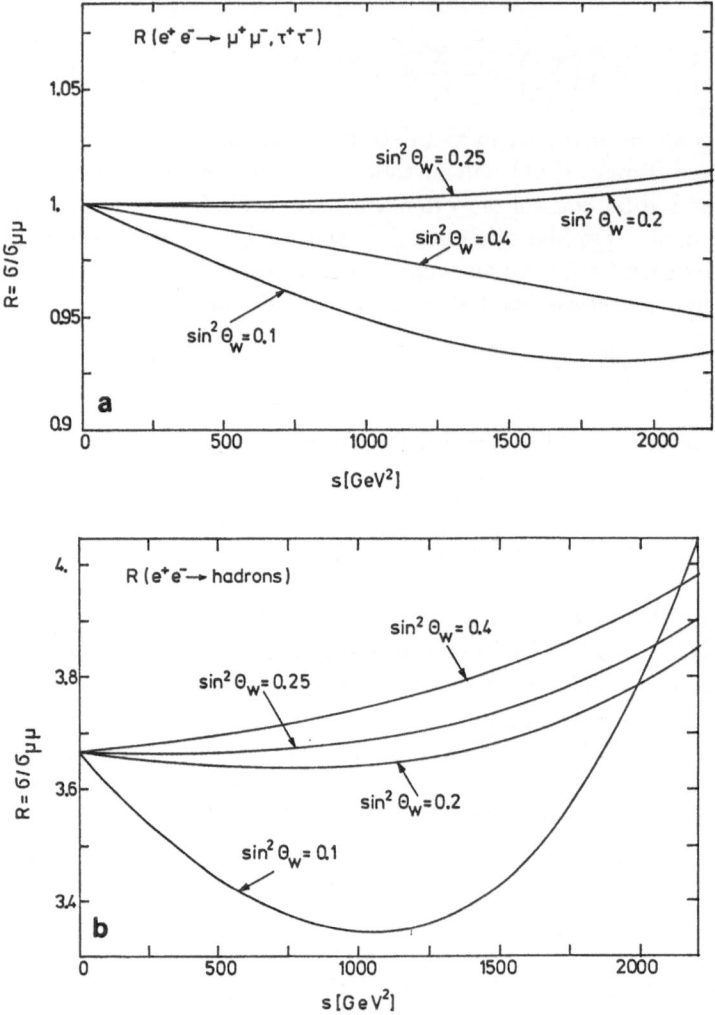

Fig. 3.4. Total cross sections for $e^+e^- \to$ leptons (a) and for $e^+e^- \to$ quarks (b) for various values of the Weinberg angle

small for leptonic final states in the PEP/PETRA energy range for values of $\sin^2 \theta_W$ around 0.25, while the effect amounts to about 6 % at the top end of PETRA for the total cross section into hadrons.

3.1.3 Polarisation

With unpolarised beams, the polarisation $P_f(\theta)$ of the final state fermion is defined as

$$P_f(\theta) = \frac{\frac{d\sigma}{d\Omega}(f_{\rm R}) - \frac{d\sigma}{d\Omega}(f_{\rm L})}{\frac{d\sigma}{d\Omega}(f_{\rm R}) + \frac{d\sigma}{d\Omega}(f_{\rm L})} \tag{3.26}$$

where $d\sigma(f_i)/d\Omega$ is the differential cross section for producing a fermion with handedness i. Using the helicity amplitudes (3.12, 13) etc., these differential cross sections are given by

$$\frac{4s}{\alpha^2}\frac{d\sigma}{d\Omega}(f_{\rm L}) = \frac{1}{4}\left[|Q_f - 8c_{\rm L}(e)c_{\rm L}(f)g(s)|^2(1+\cos\theta)^2 \right.$$
$$\left. + |Q_f - 8c_{\rm R}(e)c_{\rm L}(f)g(s)|^2(1-\cos\theta)^2\right]$$

$$\frac{4s}{\alpha^2}\frac{d\sigma}{d\Omega}(f_{\rm R}) = \frac{1}{4}\left[|Q_f - 8c_{\rm L}(e)c_{\rm R}(f)g(s)|^2(1-\cos\theta)^2 \right.$$
$$\left. + |Q_f - 8c_{\rm R}(e)c_{\rm R}(f)g(s)|^2(1+\cos\theta)^2\right] .$$

(3.27)

Inserting (3.27) into (3.26) yields

$$P_f(\theta) = \frac{F_{\rm S}(1+\cos^2\theta) + 2F_{\rm A}\cos\theta}{C_{\rm S}(1+\cos^2\theta) + 2C_{\rm A}\cos\theta} .$$

(3.28)

For the coefficients $F_{\rm S}$ and $F_{\rm A}$ one obtains, using the coupling constants v_i and a_i as defined in (2.36)

$$F_{\rm S} = Q_f v_e a_f {\rm Re}\{g(s)\} - \frac{1}{2}(v_e^2 + a_e^2)v_f a_f |g(s)|^2$$

$$F_{\rm A} = Q_f v_f a_e {\rm Re}\{g(s)\} - \frac{1}{2}(v_f^2 + a_f^2)v_e a_e |g(s)|^2 .$$

(3.29)

An approximate, but transparent expression for the polarisation follows from retaining only the interference contributions, neglecting the asymmetric part $C_{\rm A}$ in the differential cross section and all weak contributions to $C_{\rm S}$:

$$P_f(\theta) = {\rm Re}\{g(s)\}\left(v_e \frac{a_f}{Q_f} + a_e \frac{v_f}{Q_f}\frac{2\cos\theta}{1+\cos^2\theta}\right) .$$

(3.30)

The polarisation thus contains a part independent of $\cos\theta$, which is small due to the factor v_e, and a part antisymmetric in $\cos\theta$, which, due to the factor v_f/Q_f, is large for quarks and small for leptons. The polarisation expected for leptons and quarks is shown in Fig. 3.5 for a representative centre of mass energy of $\sqrt{s} = 34.5$ GeV.

It should be noted that the average polarisation $\langle P_f\rangle$ over all angles, due to the $\cos\theta$ term, is close to zero as long as purely weak terms can be neglected. Integrating (3.30) over θ leads to the average polarisation

$$\langle P_f\rangle = {\rm Re}\{g(s)\}\frac{v_e\, a_f}{Q_f} ,$$

(3.31)

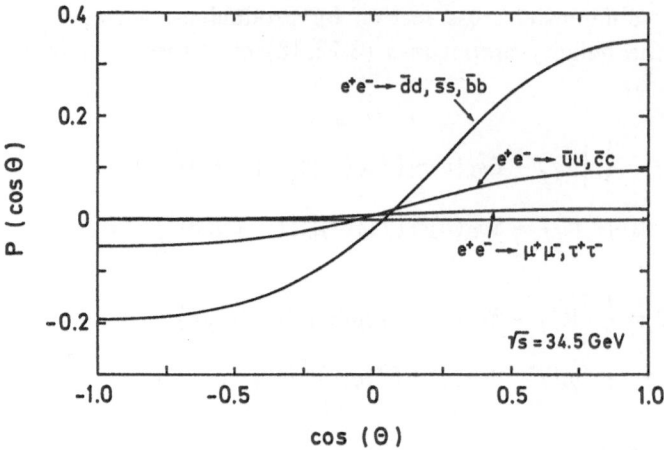

Fig. 3.5. Polarisation as function of the scattering angle, predicted by the standard theory. The curves are for $\sin^2 \theta_W = 0.23$ and $M_Z = 92$ GeV

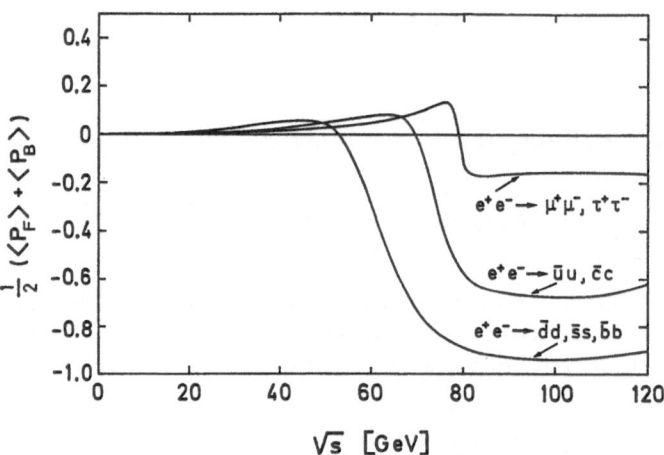

Fig. 3.6. Average polarisation as function of the centre of mass energy, as expected in the standard theory. The values $\sin^2 \theta_W = 0.23$ and $M_Z = 92$ GeV have been used to compute the curves

which projects out the constant part in (3.30), proportional to the weak vector charge v_e of the electron and to the axial charge a_f of the final state fermion. Figure 3.6 displays $\langle P_f \rangle$ as a function of the centre of mass energy for final state leptons and quarks. For energies well below the Z^0 these average polarisation are seen to be negligible.

In order to become sensitive to the odd term in (3.30) and thus to the weak vector charge of the final state fermion, one may define, similar to the charge asymmetry, a forward-backward polarisation asymmetry $A(P_f)$

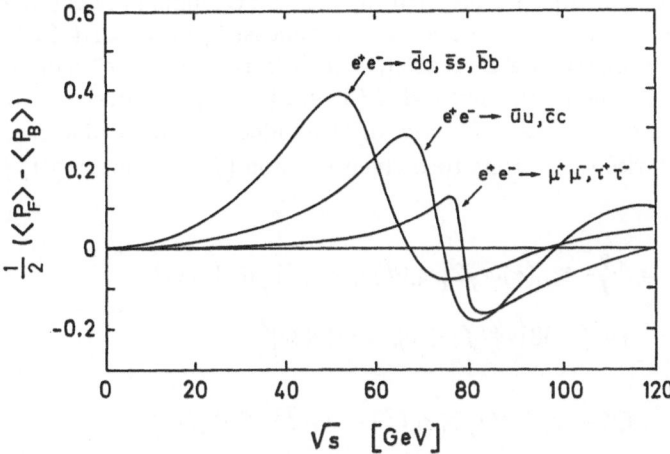

Fig. 3.7. Polarisation asymmetry as function of the centre of mass energy, as expected in the standard theory. The values $\sin^2 \theta_W = 0.23$ and $M_Z = 92$ GeV have been used to compute the curves

$$A(P_f) = \frac{1}{2}\Big(\langle P_F \rangle - \langle P_B \rangle\Big) \tag{3.32}$$

where

$$\langle P_F \rangle = \frac{\int\limits_0^1 P_f(\theta)\, d\cos\theta}{\int\limits_0^1 d\cos\theta} \quad , \quad \langle P_B \rangle = \frac{\int\limits_{-1}^0 P_f(\theta)\, d\cos\theta}{\int\limits_{-1}^0 d\cos\theta} \; .$$

Integrating (3.30) over hemispheres in θ then leads to the approximate polarisation asymmetry, neglecting purely weak terms

$$A(P_f) = \text{Re}\{g(s)\}\, \frac{a_e\, v_f}{Q_f} \ln 2 \; . \tag{3.33}$$

Predictions for the polarisation asymmetry in the standard theory as a function of the centre of mass energy are shown in Fig. 3.7. Here, as also for Fig. 3.6, the full expression for $P_f(\theta)$ (see (3.28)) has been used. At PEP/PETRA energies the expected effects in the standard theory are quite small, of order of a percent for μ and τ, but still it seems to be worthwhile to establish experimentally that the polarisations indeed follow the predictions. Furthermore, such measurements are one of the few tools at present (awaiting polarized electron beams at SLC and, potentially, at LEP) to measure the vector charges of the fermions, due to the large multipliers of v_f in (3.33).

For completeness, expressions for quantities measurable with a longitudinal polarisation of the e^+ or e^- beams will be given. Such experiments do not exist

29

at present, but polarised e^- beams are planned for the SLAC single pass collider SLC, where the electrons can be extracted from a polarised source. At LEP, even more ambitious experiments are envisaged, with *both* the e^+ and e^- beams polarised. Using again the helicity amplitudes (3.12, 13) etc., one derives the expressions for the cross section for left- and right-handed incoming electrons to annihilate with unpolarised positrons to a given fermion (f)–antifermion $(\bar{f})$ pair

$$
\frac{3s}{4\pi\alpha^2}\sigma(e_{\rm L}^-) = Q_f^2 - 4c_{\rm L}(e)Q_f 2\Big(c_{\rm L}(f) + c_{\rm R}(f)\Big)\text{Re}\{g(s)\}
$$
$$
+ 4c_{\rm L}^2(e)8\Big(c_{\rm L}^2(f) + c_{\rm R}^2(f)\Big)|g(s)|^2
$$
$$
\frac{3s}{4\pi\alpha^2}\sigma(e_{\rm R}^-) = Q_f^2 - 4c_{\rm R}(e)Q_f 2\Big(c_{\rm L}(f) + c_{\rm R}(f)\Big)\text{Re}\{g(s)\} \tag{3.34}
$$
$$
+ 4c_{\rm R}^2(e)8\Big(c_{\rm L}^2(f) + c_{\rm R}^2(f)\Big)|g(s)|^2 \ .
$$

The cross section asymmetry $A(\sigma)$, defined as

$$
A(\sigma) = 2\frac{\sigma(e_{\rm R}^-) - \sigma(e_{\rm L}^-)}{\sigma(e_{\rm R}^-) + \sigma(e_{\rm L}^-)} \tag{3.35}
$$

for completely polarised e^- beams (note that $\sigma(e_{\rm R}^-) + \sigma(e_{\rm L}^-) = 2\sigma$), is then given by

$$
A(\sigma) = \frac{a_e v_f Q_f \text{Re}\{g(s)\} - \frac{1}{2}v_e a_e(v_f^2 + a_f^2)|g(s)|^2}{Q_f^2 - Q_f v_e v_f \text{Re}\{g(s)\} + \frac{1}{4}(v_e^2 + a_e^2)(v_f^2 + a_f^2)|g(s)|^2} \ . \tag{3.36}
$$

In case of beam polarisations less than unity the observable asymmetry reduces in proportion to the actual beam polarisation. Making the usual approximations valid in the PEP/PETRA energy range the cross section asymmetry for annihilation of completely polarised electrons with unpolarised positrons into hadrons reads

$$
A(\sigma_{\rm hadr}) \approx a_e \ \text{Re}\{g(s)\} \frac{\sum_q Q_q v_q}{\sum_q Q_q^2} \ , \tag{3.37}
$$

which amounts to $A(\sigma_{\rm hadr}) \sim 0.1$ for $\sqrt{s} \approx 45$ GeV, an effect comparable in size to the charge asymmetry. The energy dependence of the cross section asymmetry is shown in Fig. 3.8, using the full formula (3.36).

3.1.4 Bhabha Scattering

Elastic scattering of electrons and positrons (Bhabha scattering), although the most elementary process in e^+e^- reactions, shows a rather complicated dependence on the scattering angle θ and, in particular, on the weak coupling constants

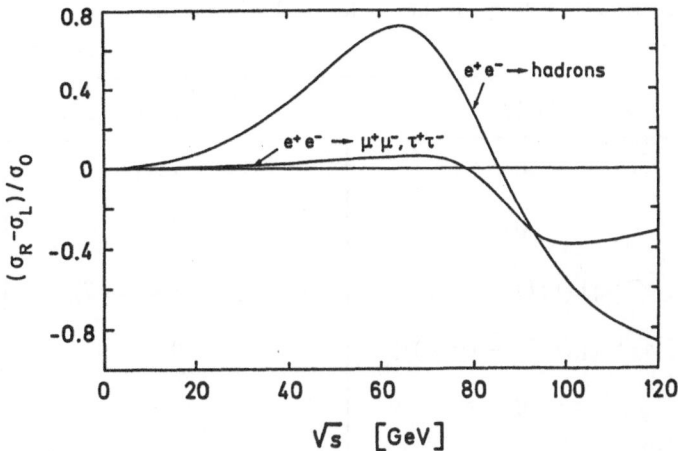

Fig. 3.8. Cross section asymmetry for a longitudinally completely polarised electron beam as function of the centre of mass energy, as expected in the standard theory. The values $\sin^2 \theta_W = 0.23$ and $M_Z = 92$ GeV have been used to compute the curves

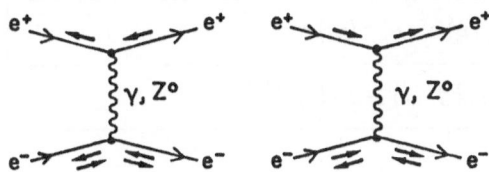

Fig. 3.9. t-channel diagrams in lowest order for Bhabha scattering. The arrows denote the possible helicity configurations of the scattering partners

v and a. This is due to the identity of initial and final state leptons which entails t-channel γ, Z^0 exchange in addition to the s-channel annihilation diagrams of Figs. 3.1,2 with amplitudes given in Table 3.1. In addition to these amplitudes there are four t-channel helicity amplitudes (for the corresponding Feynman diagrams see Fig. 3.9) given by the requirement that the $e^{\pm}$ helicities are conserved in the t-channel process.

The complete set of helicity amplitudes for s- and t channel is given in Table 3.2. Note that there are now helicity configurations possible through t channel exchange which were forbidden in s channel exchange, such as initial state fermions with identical handedness ($e_R^- e_R^+$ or $e_L^- e_L^+$). One further comment is in order to precisely define the meaning of the indices i, j in Table 3.2 as opposed to those given in Table 3.1: The index $i(j)$ in Table 3.1 referred to the handedness of the initial (final) state fermion current. This is also true for the s channel Bhabha amplitudes of Table 3.2. In the t channel, however, the fermion currents connect the initial and final states, which makes it is more appropriate to talk about "upper" and "lower" fermion currents, as depicted in Fig. 3.9. With this terminology the indices $i(j)$ refer to the handedness of the lower (upper) *fermion* current, so that, e.g., the upper current in the first diagram of Fig. 3.9 (a left-handed positron) is to be considered a right-handed fermion current.

Table 3.2. Lowest order electroweak helicity amplitudes for Bhabha scattering (s and t channel)

Helicity config.	$F_s\,(ij)$	$F_t\,(ij)$
$e_L^- e_L^+ \to e_L^- e_L^+$	0	$t^{-1}\varepsilon_{LR}(t)$
$e_L^- e_L^+ \to e_R^- e_R^+$	0	0
$e_L^- e_R^+ \to e_L^- e_R^+$	$s^{-1}\varepsilon_{LL}(s)\,(1+\cos\theta)/2$	$t^{-1}\varepsilon_{LL}(t)\,(1+\cos\theta)/2$
$e_L^- e_R^+ \to e_R^- e_L^+$	$s^{-1}\varepsilon_{LR}(s)\,(1-\cos\theta)/2$	0
$e_R^- e_R^+ \to e_R^- e_R^+$	0	$t^{-1}\varepsilon_{RL}(t)$
$e_R^- e_R^+ \to e_L^- e_L^+$	0	0
$e_R^- e_L^+ \to e_R^- e_L^+$	$s^{-1}\varepsilon_{RR}(s)\,(1+\cos\theta)/2$	$t^{-1}\varepsilon_{RR}(t)\,(1+\cos\theta)/2$
$e_R^- e_L^+ \to e_L^- e_R^+$	$s^{-1}\varepsilon_{RL}(s)\,(1-\cos\theta)/2$	0
with $\varepsilon_{ij}\,(w) = -1 - 8\,c_i\,(e)\,c_j\,(e)\,g\,(w)\,;\quad w = s,\,t$		
$d\sigma/d\Omega = \frac{\alpha^2 s}{4}\,\sum_{ij}\,|F_s\,(ij) + F_t\,(ij)|^2$		

Adding the s- and t-channel contributions for a given helicity state, and summing over the squared amplitudes thus obtained, leads to the following differential cross section (unpolarised beams) including electroweak effects:

$$\frac{4s}{\alpha^2}\frac{d\sigma}{d\Omega}(e^+e^- \to e^+e^-) = (1+\cos^2\theta)\,C_S + 2C_A\cos\theta - \frac{2(1+\cos\theta)^2}{1-\cos\theta}\,C_{\text{int}}$$
$$+ \frac{8}{(1-\cos\theta)^2}\,C_{t_1} + \frac{2(1+\cos\theta)^2}{(1-\cos\theta)^2}\,C_{t_2}\,.$$
(3.38)

C_S and C_A were given in (3.15) and arise from the s-channel diagrams contributing to Bhabha scattering. The coefficients multiplying the t-channel terms (C_{t_i}) and the s-t-channel interference term (C_{int}) are given by

$$C_{\text{int}} = 1 + \frac{1}{2}\,(v_e^2 + a_e^2)\,(\text{Re}\{g(s)\} + \text{Re}\{g(t)\})$$
$$+ \frac{1}{4}\,(v_e^4 + 6v_e^2 a_e^2 + a_e^4)\cdot\text{Re}\{g^*(s)g(t)\}\,,$$

(3.39)

$$C_{t_1} = 1 + (v_e^2 - a_e^2)\,\text{Re}\{g(t)\} + \frac{1}{4}(v_e^2 - a_e^2)^2|g(t)|^2\,,$$

$$C_{t_2} = 1 + (v_e^2 + a_e^2)\,\text{Re}\{g(t)\} + \frac{1}{4}\,(v_e^4 + 6v_e^2 a_e^2 + a_e^4)|g(t)|^2\,.$$

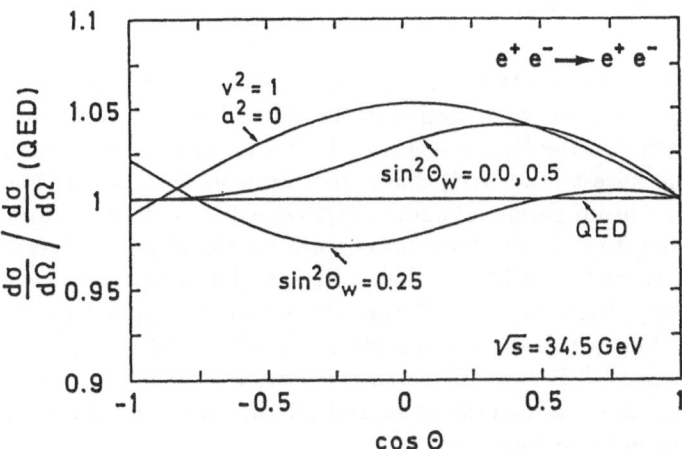

Fig. 3.10. Ratio of the Bhabha cross section as given by the standard theory to the lowest order QED expectation for various values of the Weinberg angle, and for $v^2 = 1, a^2 = 0$, which cannot occur in the standard theory

The propagator function $g(w)$, $w = s$, t was defined in $(3.9, 10)$. For the differential cross section for Bhabha scattering in lowest order QED, the Z^0 propagator terms $g(s)$ and $g(t)$ are set to zero thus yielding

$$\frac{d\sigma}{d\Omega}(e^+e^- \rightarrow e^+e^-)_{\mathrm{QED}} = \frac{\alpha^2}{4s} \frac{(3 + \cos^2\theta)^2}{(1 - \cos\theta)^2} . \tag{3.40}$$

From (3.40) it is evident that the Bhabha cross section is dominated by the t-channel exchange diagrams leading to a large cross section at small angles. Here also the contributions from weak effects vanish $(g(t) \rightarrow 0$ for $t \rightarrow 0)$. The backward hemisphere, on the other hand, is sensitive to weak effects.

Figure 3.10 shows the ratio of the electroweak cross section (3.38) to the lowest order QED cross section (3.40) for various values of the Weinberg angle as a function of the scattering angle θ at a center of mass energy of $\sqrt{s} = 34.5$ GeV. Also given in the figure is the expectation for $v^2 = 1$, $a^2 = 0$, which cannot occur within the standard model, but is one of the possible solutions for the electroweak charges from neutrino-electron scattering (see Chap. 4). For $\sin^2\theta_W \approx 0.23$, very small deviations from QED are expected, amounting to $\sim -3\%$ in the backward hemisphere. The reason for these small effects are cancellations in the terms multiplied by C_{t_1} and C_{t_2} (see (3.39)): With v_e being almost zero for values of $\sin^2\theta_W$ around $1/4$, the weak corrections to the two t channel coefficients are proportional to $-a_e^2$ and $+a_e^2$, respectively. Furthermore the weak contributions to the interference term C_{int} cancel due to $g(s) \approx -g(t)$.

3.2 Radiative Corrections

When the standard theory is confronted with experiment, extreme care has to be taken on either side to control the systematics to the desired level of precision. Concerning the systematics on the theoretical side, higher order QED and weak diagrams will add to the lowest order (α^2) diagrams shown in Figs. 3.1,2. For quark final states also quantum chromodynamics (QCD) will contribute. Thus, in general, the expressions given in the previous chapter for the physical observables will have to be modified in order to allow a meaningful comparison with experiment. Alternatively, these "radiative" corrections can be applied to the data, leading to observables which display the physics produced by the neutral current as discussed in Sect. 3.1. As will be seen below, the radiative corrections depend largely on the experimental setup and on the particular methods of extracting the data from the experiment.

In order to isolate the weak effects and to enable comparison of data from different experiments with each other and with the theoretical predictions, the higher order contributions are customarily estimated and subtracted from a given measured quantity. In the following the corrections up to order α^3 will be considered for which extensive calculations exist in the literature. Pure QED and purely weak corrections are discussed seperately.

The next to lowest order pure QED contributions to the Born terms of Figs. 3.1,2 may be divided into two classes: a) Adding (fermion) loops to the internal lines, known as vacuum polarisation terms, and b) adding a photon line in all possible ways to the external lines. Let us first discuss these two classes of higher order corrections for the QED Born term of Fig. 3.1. The vacuum polarisation diagrams are symbolically depicted in Fig. 3.11.

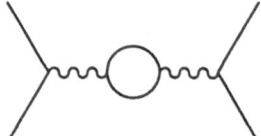

Fig. 3.11. Vacuum polarisation diagrams contributing to the photon propagator. The internal loop represents an arbitrary fermion

As is well known, the contributions from the vacuum polarisation diagrams lead to the notion of an effective fermion charge, which depends on Q^2, the momentum transfer squared involved in the scattering process. In colloquial terms, the charge polarises the surrounding vacuum filled with virtual fermion-antifermion pairs and is thus screened to a lesser degree the closer the distance, i.e. the higher the momentum transfer, between the scattering particles. The effect of the vacuum polarisation diagrams is therefore a modification of the photon propagator in the QED Born term, which can be absorbed in the definition of running fine structure constant $\alpha(Q^2)$. In the modified minimal substraction ($\overline{\text{MS}}$) scheme, Marciano (1979) gives the following expression

$$\alpha^{-1}(Q^2) = \alpha^{-1} - \frac{1}{3\pi} \sum_f Q_f^2 \ln\left(\frac{Q^2}{m_f^2}\right) + O(\alpha). \qquad (3.41)$$

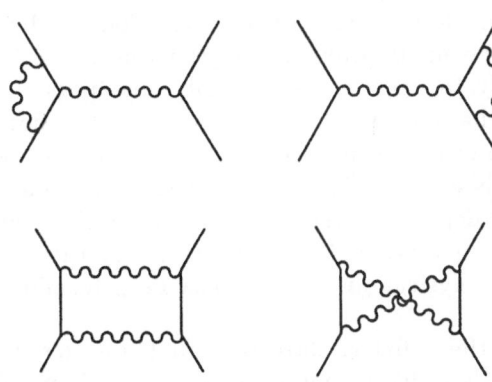

Fig. 3.12. QED vertex correction and box diagrams

Here, α is the usual fine structure constant derived from low energy ($Q^2 \sim m_e^2$) experiments. The sum is over all fermions with charge Q_f (a colour factor of three must be included for quarks) and masses $m_f^2 < Q^2$. Note that the above expression holds for the continuum region of the $f\bar{f}$ final states, away from the poles such as the $q\bar{q}$ vector meson resonances $\rho, \phi, J/\psi, \Upsilon$ etc.. For a numerical evaluation of (3.41) effective masses of 0.1 GeV may be used for u, d, and s quarks, consistent with a dispersive analysis of $e^+e^- \to$ hadrons by Berends and Komen (1976). It is instructive to recall that, by virtue of (3.41), the effective α has increased from $\sim 1/137$ to $\sim 1/129$ at a typical PETRA energy of 35 GeV.

The second class of corrections to the lowest order QED Born term, i.e. addition of an extra photon to the external lines, leads to two different sets of diagrams. The virtual corrections, with the additional photon as an internal line, are represented by the diagrams in Fig. 3.12 (fermion self-energy diagrams are absent due to the on-mass-shell renormalisation scheme traditionally used in QED calculations). Interference of these diagrams with the Born term diagram of Fig. 3.1 leads to $O(\alpha^3)$ corrections. On the other hand, the additional photon may be added as an external line to the Born term as shown in the diagrams of Fig. 3.13. These bremsstrahlung contributions are of order α^3. It should be

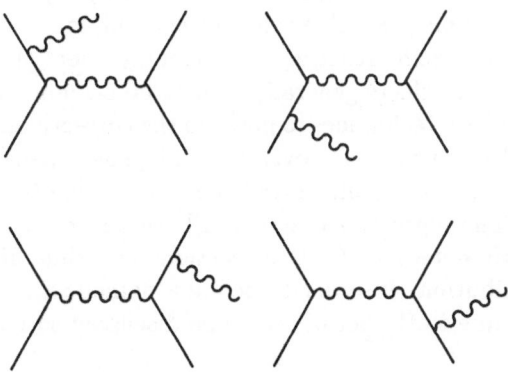

Fig. 3.13. QED diagrams to order α^3 with an external photon line

35

mentioned here that both contributions derived from vertex correction and box diagrams of Fig. 3.12 and those of the bremsstrahlung diagrams of Fig. 3.13 are infrared divergent, when taken by themselves. Technically one splits the bremsstrahlung diagrams in a finite hard part (photon observed in the final state) and a divergent soft part which exactly cancels the divergent part of the virtual contributions (see e.g. Berends et al., 1973). As the detection of hard bremsstrahlung photons will strongly depend on the detector capabilities, the relative contributions of the diagrams in Figs. 3.12, 13, and therefore the size of the radiative correction for a given physical observable, must be determined individually for each experiment.

In order to illustrate the effects of the radiative corrections first consider the total cross section for $e^+e^- \to f\bar{f}$. Here, hard photon emission in the initial state will lead to an increase of the number of observed events: After radiating a photon, the electron positron annihilation occurs at a reduced centre of mass energy with a correspondingly higher cross section, due to its $1/s$ dependence. If, on the other hand, events with detected photons radiated off the initial or final state fermions are rejected, the vacuum polarisation diagram Fig. 3.11 will give the dominant correction.

The charge asymmetry will receive important $O(\alpha^3)$ corrections depending on the admitted kinematical region of the radiative photon. The interference between the Born term and box diagrams of Fig. 3.12, supplemented with the soft part of the interference between initial and final state bremsstrahlung of Fig. 3.13, leads to a positive asymmetry, while the interference of latter diagrams describing hard bremsstrahlung gives negative contributions. A convenient way to control the available phase space for the radiative photon, and thereby to define the event sample of the reaction $e^+e^- \to f\bar{f}$, is to apply a cut in the acollinearity angle ξ given by the momentum vectors $\vec{P}_{\bar{f},f}$ of the outgoing fermions

$$\xi = \cos^{-1}(-\vec{P}_{\bar{f}} \cdot \vec{P}_f) . \qquad (3.42)$$

Admitting, e.g. in the reaction $e^+e^- \to \mu^+\mu^-$, only nearly collinear μ pairs by the requirement $\xi < O(10°)$ will lead to typical overall corrections of $\sim +25\%$ for the cross section around 35 GeV and of $\sim +1.5\%$ for the charge asymmetry from $O(\alpha^3)$ QED. The experimentally observed charge asymmetry will differ by that amount from the negative values expected in the standard theory. In practice, detector acceptance and resolution functions, as well as sometimes complex cuts necessary to purify event samples for a specific reaction, will strongly affect the radiative corrections. These experimental effects generally cannot be calculated analytically and one resorts to Monte Carlo techniques to perform the corrections. Computer programs generating final state particles over the full phase space, properly taking into account $O(\alpha^3)$ matrix elements, have become available (see references in the following sections). These programs serve as a basis to simulate the reaction under study in a realistic detector. QED corrections can thus be calculated for any experimental distribution. In order to define a nomenclature for later applications the corrections due to the set of diagrams discussed so far will be labelled "reduced QED".

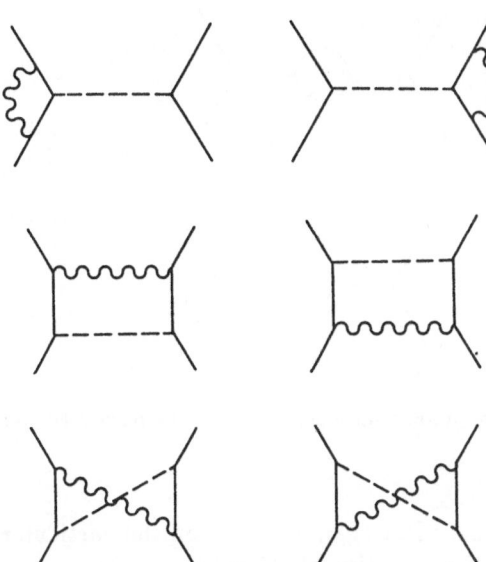

Fig. 3.14. Virtual $\gamma - Z^0$ diagrams, the lines denote the Z^0

The radiative corrections of the QED Born term discussed so far also apply for the Z^0 exchange diagram of Fig. 3.2. In addtion, weak loop corrections have to be considered. The corresponding lowest order diagrams are shown in Figs. 3.14–16. Combining these additional diagrams with those of Figs. 3.12,13 and the two Born terms of Figs. 3.1,2 to form amplitudes of $O(\alpha^3)$ leads again to cancellations of the infrared divergencies arising from the virtual and soft bremsstrahlung contributions, similar to the gauge invariant subset of pure QED diagrams discussed above. The sum of corrections resulting from reduced QED and the radiative corrections to the Z^0 Born term are labelled "full QED". As to the charge asymmetry, a further reduction in magnitude is expected by the full QED calculation over the reduced QED correction. Since initial state radiation will lower the centre of mass energy available for the annihilation process, the Z^0 propagator term (see (3.11)) will consequently lead to a smaller effect in the

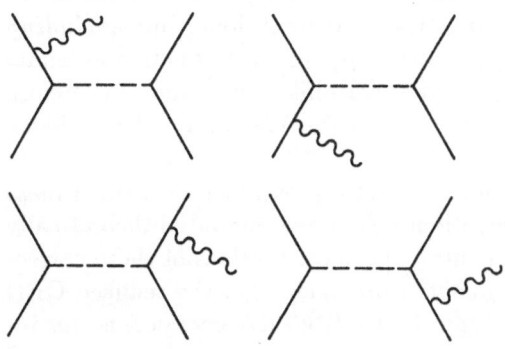

Fig. 3.15. Feynman diagrams for Z^0 exchange with an additional external photon line

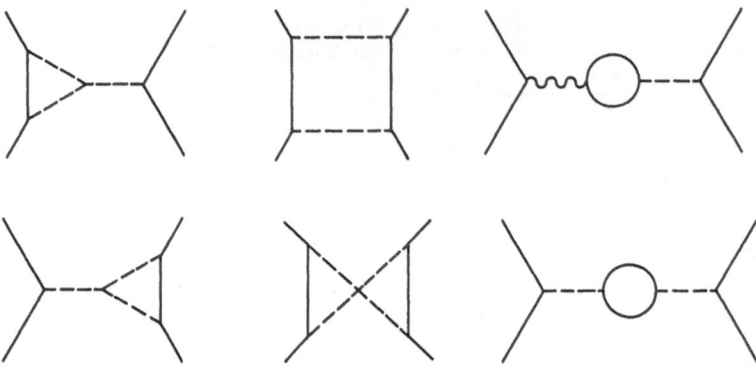

Fig. 3.16. Virtual weak loop diagrams. The dashed lines denote the heavy bosons $W^\pm$ or Z^0

charge asymmetry. Clearly, the size of the effect again depends on the particular experimental conditions for measuring the radiative photons.

The standard theory is renormalisable and therefore higher order weak loop corrections can be calculated. Some typical diagrams encountered are shown in Fig. 3.16. The corresponding virtual corrections to third order in the coupling strengths are generated by interference of the loop terms of Fig. 3.16 with the Born terms of Figs. 3.1,2. Due to the nonzero mass of the weak bosons these corrections do not exhibit any infrared divergencies. They are furthermore independent of the experimental capabilities to detect photons in the final state.

The relative magnitudes of the Z^0 Born term and the electroweak corrections calculated to a limited order depend, in general, on the renormalisation scheme used. Two such schemes are most commonly used, as already indicated by the parametrisations of the Z^0 propagator in (3.9, 10): For (3.9) the renormalised parameters are α, M_Z and $\sin^2\theta_W$ or M_W (Glashow scheme), whereas for (3.10) α, G_F and M_Z are used (Weinberg-Salam scheme). A large amount of work has been done to calculate the electroweak corrections in both schemes (see, e.g., Passarino and Veltman, 1979; Wetzel, 1983; Böhm and Hollik, 1984; Brown et al., 1984). The full QED calculation, using the Glashow scheme, has been carried out by Berends et al. (1982), the corresponding Monte-Carlo program for the reaction $e^+e^- \to \bar{f}f(f \neq e)$ is available (Jadach, 1983). Whereas in the Glashow scheme the electroweak (radiative and weak loop, but *excluding* pure QED) correction to the charge asymmetry in $\mu(\tau)$ pair production amounts to about -0.1% at PETRA energies (Böhm and Hollik, 1984) and is therefore entirely negligible, the corresponding correction in the Weinberg-Salam scheme is of order 1% (Wetzel, 1983).

Table 3.3 summarises these predictions for two representative centre of mass energies, The electroweak radiative corrections have been computed theoretically by choosing "reasonable" experimental cuts. The precise values of these corrections will depend on the specific experiment. Note again that the reduced QED correction, which is about $+1.5\%$ at typical PEP/PETRA energies, is *not* in-

Table 3.3. Electroweak corrections for the $\mu(\tau)$ charge asymmetry A_{FB} in percent for two different renormalisation schemes (see text). The values $M_Z = 92$ GeV and $\sin^2 \theta_{\mathrm{W}} = 0.23$ are used.

$\sqrt{s}$[GeV]	Born	rad.	weak loops	sum	renorm. scheme
34.5	−8.62	+0.59	−0.69	−8.72	Glashow
	−9.29	+0.56	−0.04	−8.69	W – S
44.0	−15.50	+1.11	−1.23	−15.62	Glashow
	−16.68	+1.00	−0.10	−15.58	W – S

cluded in the Table. The first column gives the Born term values, the second and third columns the radiative and weak loop corrections to the Z^0 Born term, respectively. The fourth column is the sum over the preceeding columns, i.e. the prediction for the charge asymmetry in the standard theory at the one-loop-level.

Although differing substantially at the Born level, the one-loop corrected values agree nicely for both schemes. In addition, the one-loop correction to the Born term in the Glashow scheme is well below present-day experimental accuracy. The tiny differences observed for the total electroweak correction (sum) in both schemes can be attributed to higher order corrections beyond the one-loop level. Since traditionally differential cross sections have been corrected for pure ("reduced") QED only, the simplest method to compare measured charge asymmetries with the standard theory thus seems the following: Only apply reduced QED corrections to the data and compare to the lowest order prediction in the Glashow scheme, using the parametrisation of the propagator $g(s)$ according to (3.9). In fact, all data on $\mu(\tau)$ asymmetries presented in this review are treated that way. On the other hand, as data may become more accurate in the future, especially at the "Z^0 machines", the better way will be to apply the full QED correction to the data and to correct for the weak loop contributions. It is understood that this method of correcting the data is iterative, since the corrections depend on parameters (mass of the Z^0, value of the Weinberg angle) one hopes to determine from the data.

3.3 Experiments

Two e^+e^- colliding beam machines have been operating in the past years with energies high enough to study electroweak effects: PEP (*Positron-Electron-Project*) at the Stanford Linear Accelerator Center SLAC, Stanford, USA and PETRA (*Positron-Elektron-Tandem-Ringbeschleuniger-Anlage*) at the Deutsches Elektronen Synchrotron (DESY), Hamburg, Germany. Both storage rings have terminated their physics program by the end of 1986. While PEP will still be usable for e^+e^- physics in the future, PETRA is in the process of being modified to serve as pre-accelerator for the HERA electron-proton machine.

PEP started operation in 1980, two years after PETRA, and has almost exclusively been running at a fixed centre of mass energy of 29 GeV. At this energy PEP operates at high luminosity. The luminosity $\mathcal{L}$ of a colliding beam machine is a measure of the event rates N obtained for a given cross section σ and is defined by

$$N = \mathcal{L}\sigma . \qquad (3.43)$$

Since the beginning of 1983 an average integrated luminosity of $\sim 1\,\mathrm{pb}^{-1}$, corresponding to an equivalent of $\sim 100\ e^+e^- \to \mu^+\mu^-$ events, has been delivered per day. The total integrated luminosity by the time of interrupting the PEP operation for the construction of the SLC was close to $\sim 300\,\mathrm{pb}^{-1}$ for the running experiments.

PETRA, on the other hand, has collected data over a wide range of center of mass energies, between 14 GeV up to a maximum of 46.78 GeV. Most of the time was spent searching for the top quark, or any other new threshold, by scanning the region above 30 GeV in narrow energy steps with a statistics of $\sim 50\,\mathrm{nb}^{-1}$ per point and experiment. The step size varied between 20 MeV and 30 MeV (above $\sqrt{s} \sim 38$ GeV), and was chosen to be smaller than the machine energy spread in order to continuously cover the energy range. The bulk of data, equivalent to $\sim 60\,\mathrm{pb}^{-1}$ per experiment, was obtained at energies around 34.5 GeV. Although the luminosity of PETRA was constantly improved the average daily integrated luminosity, due to the densely packed RF cavities required for the high energy running and the less efficient scanning operation mode, has been substantially smaller compared to PEP. In the high energy region, with an average of $\sqrt{s} \sim$ 44 GeV, about $40\,\mathrm{pb}^{-1}$ per experiment have been collected by the end of 1985. The smaller integrated luminosity is mostly counterbalanced, however, by the larger centre of mass energy $\sqrt{s}$, since the size of the electroweak effects is growing linearly with s.

In its last year of running PETRA was set to an energy of 35 GeV. With the known lack of "new" physics at the top end of the machine, this seemed a good compromise between the high energies desirable to study electroweak effects and a substantial increase in the total number of events recorded. At this setting the PETRA experiments have collected an equivalent of about 90 pb^{-1} each.

Recently, an new e^+e^- machine has come to operation at KEK, Japan. TRISTAN, with a design energy up to almost 60 GeV, has successfully run at centre of mass energies around 54 GeV, providing first physics results at the Rochester Conference at Munich, 1988 from an equivalent of about 10 pb^{-1} for each of the three installed experiments.

Most detectors at e^+e^- machines probing a new energy regime have been designed to cover a physics range as broad as possible. Such detectors should be able to study reactions containing (few) leptons and/or (many) hadrons from the conventional sources, such as pair production of the known fermions through the s channel exchange processes discussed in Chap. 2. They should, however, also be prepared to detect exotic reactions with new classes of particles, usually characterised by large missing energy due to weakly interacting neutral reaction products. The general construction principle of such "general purpose" detectors is therefore an almost complete coverage of the solid angle ("4π detectors") with momemtum analysis of charged particles and calorimetry for electromagnetically interacting particles. As the centre of mass energiy rises, hadronic calorimeters will become more and more important.

Let us first discuss the requirements for the magnetic detector components. An important feature of a detector for electroweak physics is the possibility to determine the charge sign (and momentum) of very energetic outgoing leptons without any bias. The charge sign is crucial for the measurement of the angular asymmetry, e.g. in μ pair production. The charge sign of the muons in this reaction, which have very large momenta and therefore lead to almost straight tracks in the detector, is derived from a curvature measurement. Due to the finite resolution of the tracking devices, however, for a certain fraction of the muons a wrong charge sign will be determined. This charge confusion can, e.g., be determined experimentally by counting the fraction of like-sign muon pairs in the total two-muon sample.

The experiments also have to make sure that the relative acceptance for positively and negatively charged particles does not depend on the polar angle. In the case of a detector with a solenoidal spectrometer such a dependence could be introduced by an end-to-end azimuthal twist of the central wire chamber assembly. A possible bias in solenoid detectors can be determined, e.g., by comparing curvature distributions of cosmic ray and high energy particle tracks from both detector hemispheres. Detectors with a toroidal field, on the other hand, have a "built-in" acceptance asymmetry with respect to the charge sign, since for a given scattering angle positive muons may, depending on the magnet polarity, be bent towards the beam while negative muons are bent away from the beam. This latter asymmetry can in principle be eliminated by collecting data in about equal amounts with both magnet polarities. The inherent experimental subtleties of such magnetic field arrangements, however, led most experiments to chose a "symmetric" solenoidal magnetic field configuration.

Since lepton identification is mandatory both for conventional and "new" physics, electromagnetic calorimeters are widely used, employing predominantly liquid argon or scintillator read-out techniques. Further components are added to detect muons having passed through the bulk of the detector material.

Some experiments have emphasised hadron calorimetry, thereby giving up on a large lever arm for the momentum measurement of charged particles due to reasons of economy. Most LEP/SLC experiments now under construction, by the way, have chosen to combine good momentum measurement with electromagnetic *and* hadronic calorimetry, which resulted, as a consequence of the high particle momenta and energies expected at the Z^0, in quite costly detector designs.

In the following the experiments at PEP and PETRA are briefly introduced, mentioning only a few characteristic features of the detectors and describing in more detail one of the PETRA detectors (CELLO). More information can be found in the references given below.

Five experiments have taken significant data samples at PEP:

DELCO (Kirkby, 1982) is an upgraded version of the original detector running at SPEAR (Bacino et al., 1978), with an open-geometry (Helmholtz coils) magnetic spectrometer and a threshold gas Cerenkov counter system, mainly for electron identification, covering 60% of 4π steradians, as its characteristic elements.

HRS (Ahlen et al., 1983) is built inside the former Argonne bubble chamber superconducting magnet, optimized for charged particle tracking, with a momentum resolution of $\Delta p/p = 0.001 \cdot p \,[\text{GeV}/c]$. Electromagnetic showers are detected in two layers of proportional chambers sandwiched between lead absorber plates surrounding the tracking chambers.

MAC (Ford, 1982; Weinstein, 1982) consists of a small diameter central wire detector with large angular acceptance inside a solenoidal normal conducting magnet, which is followed by gas shower chambers, a magnetized hadron calorimeter and muon chambers.

MARK II (Schindler et al., 1981) is a large diameter solenoidal spectrometer with normal-conducting coil, providing a momentum resolution of $\Delta p/p = 0.015 \cdot p \,[\text{GeV}/c]$. Outside the coil a lead-liquid argon shower counter system is installed with an angular acceptance of 65 % of 4π and an energy resolution of $\sigma(E)/E = 0.11/\sqrt{E}$, followed by iron absorbers and muon chambers covering 55 % of 4π.

TPC (Aihara et al., 1983) is the technologically most ambitious detector, featuring a time projection chamber of 2 m diameter inside a superconducting solenoid.

PETRA with only 4 interaction regions, housed the following experiments:

JADE (Bartel et al., 1979) has as its main component a pressurized pictorial drift chamber (the "jet chamber"), which provides up to 48 space points and energy loss measurements per track inside a 0.5 Tesla solenoid, covering almost 90 % of 4π. The momentum resoultion of the spectrometer is $\Delta p/p = 0.012 \cdot p \,[\text{GeV}/c]$. Electromagnetic showers are detected in a lead glass counter system, covering 90 % of 4π with an energy resolution of $\sigma(E)/E = 0.16/\sqrt{E}$. The whole detector is surrounded by sets of muon chambers interspersed between iron-loaded concrete.

MARK J (Barber et al., 1980) has emphasised hadronic calorimetry, using a lead-scintillator assembly with a thickness of 4 interaction lengths, surrounded by a toroidally magnetized muon range telescope. Charged particles are recognized by a drift tube arrangement inside the hadronic calorimeter.

TASSO (Brandelik et al., 1979, 1980) consists of a large normal-conducting solenoid magnet, equipped with cylindrical proportional and drift chambers. The tracking chambers cover 87 % of 4π with a momentum resolution of $\Delta p/p = 0.012 \cdot p \,[\text{GeV}/c]$. Outside the coil Cerenkov counters for hadron identification and liquid argon calorimeters are installed in non-overlapping angular regions, backed by iron absorbers and muon chambers. The liquid argon calorimeter system covers 44 % of 4π.

PLUTO (Berger et al., 1981; Criegee and Knies, 1982) at PETRA is the modified DORIS detector with two additional forward spectrometers. The central part consists of cylindrical proportional chambers with a momentum resolution of $\Delta p/p = 0.025 \cdot p\,[\text{GeV}/c]$ over 87 % of the full solid angle. The superconducting solenoid of 1.6 Tesla is surrounded by lead-scintillator shower counters with an energy resolution of $\sigma(E)/E = 0.35/\sqrt{E}$, followed by iron absorbers and muon chambers. Up to the summer 1983 PLUTO shared an interaction region with the CELLO detector.

CELLO (Behrend et al., 1981) was designed to identify and measure leptons, photons and hadrons with high precision over almost the entire solid angle. Electromagnetic calorimetry is strongly emphasized. The detector in its 1983 version is shown in Fig. 3.17.

The beam pipe is surrounded by two layers of drift tubes at radii of 10.5 cm and 11 cm. The central detector, consisting of 7 cylindrical drift chambers interleaved with 5 proportional chambers (outer radius 70 cm) is placed inside a thin-walled (0.47 radiation lengths) superconducting coil, 1.5 m in diameter, with a field strength of 1.3 Tesla. The central detector covers 91% of 4π. The momentum resolution obtained is $\Delta p/p = 0.013 \cdot p\,[\text{GeV}/c]$. Outside the magnet a fine grain lead liquid argon calorimeter is installed consisting of an octagon ("barrel") surrounding the coil and forward/backward modules ("end caps"). The barrel is subdivided into 16 modules housed in a single cryogenic tank. The solid angle coverage of the barrel is 86% of 4π. Each of the 16 modules samples the energy deposited by ionising particles in the liquid argon 19 times in depth. The charge is collected on lead strips of three different orientations amounting to a total of 20 radiation lengths (normal incidence). The energy resolution for electrons and photons is measured as $\sigma(E)/E = 0.13/\sqrt{E[\text{GeV}]}$, and the angular resolution is 4 mrad, assuming the interaction region as vertex. The muon detector covers 92% of 4π and consists of one layer of planar wire chambers, allowing space point reconstruction behind a hadron filter formed by the liquid argon calorimeter and 80 cm of iron. The hadron filter thickness is equivalent to ≥ 5.5 interaction lengths. The triggers are derived from wire chamber and calorimeter information incorporating a fast 2 charged particle trigger, low threshold total energy triggers and various combinations of charged and neutral particle triggers. Due to the sharing of an interaction region, CELLO could not participate in the early running period around 34.5 GeV which delivered about half of the integrated luminosity at this fixed medium energy point.

The following chapters will present the experimental information from e^+e^- experiments relevant for testing the standard theory available by the summer of 1988. Some of the results are preliminary and have only been presented at conferences. As mentionned above, data taking has stopped for the high energy e^+e^- machines PEP and PETRA, and the bulk of physics analysis concerning electroweak effects have been completed by now. A clear departure from QED of the observables proportional to the (large) axial vector coupling constants (e.g. the charge asymmetries) has been established beyond doubt. The measurements are found in complete agreement with the standard theory. To improve the accuracy substantially and to also measure the (small) vector coupling constants to the point where one becomes sensitive to deviations from the standard theory will take considerably more effort, time and, certainly, new accelerators for higher energies.

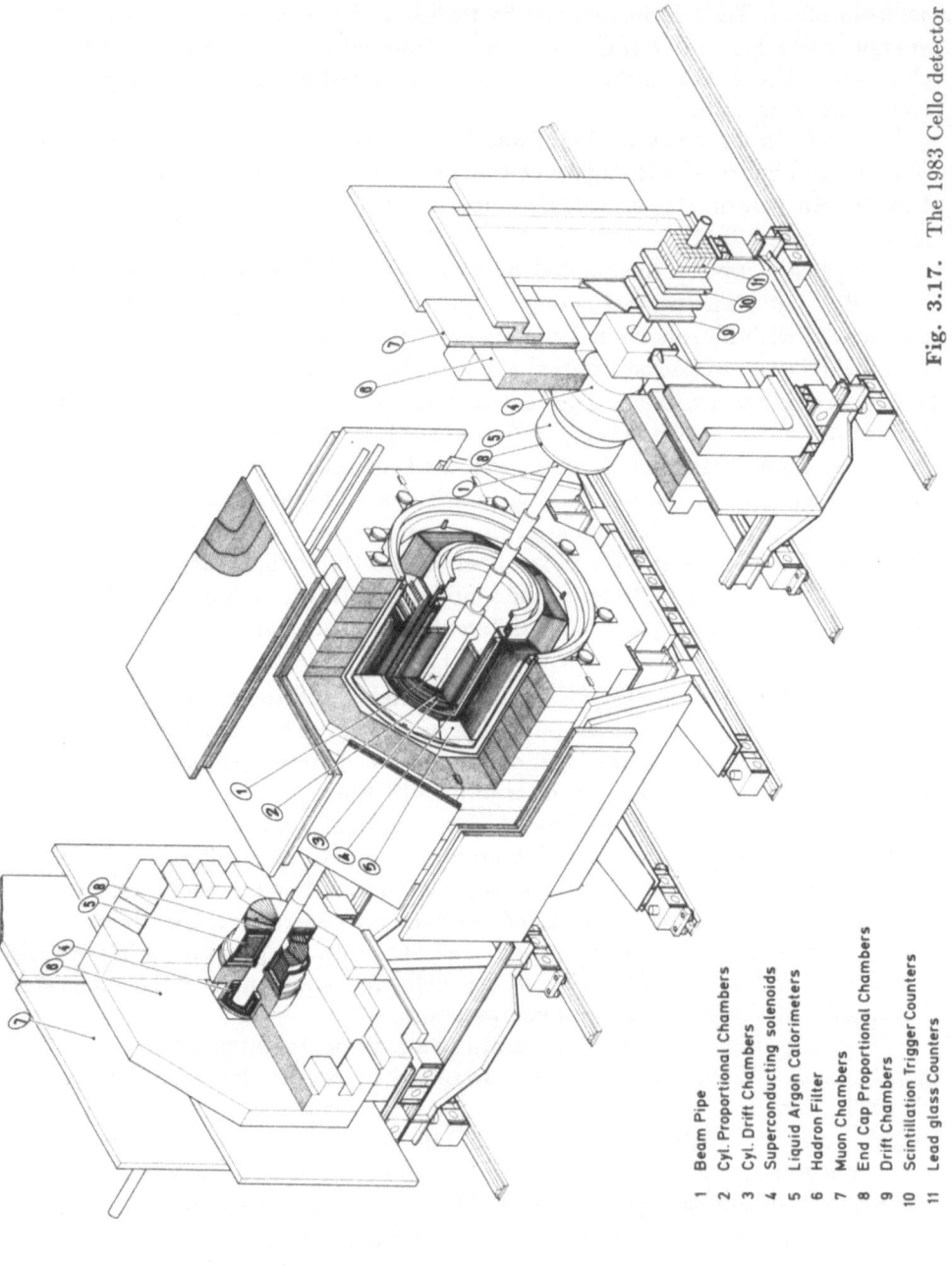

Fig. 3.17. The 1983 Cello detector

1 Beam Pipe
2 Cyl. Proportional Chambers
3 Cyl. Drift Chambers
4 Superconducting solenoids
5 Liquid Argon Calorimeters
6 Hadron Filter
7 Muon Chambers
8 End Cap Proportional Chambers
9 Drift Chambers
10 Scintillation Trigger Counters
11 Lead glass Counters

3.4 Leptonic Reactions

In the s-channel one photon exchange (see Fig. 3.1), far above the kinematical threshold, charged leptons of the different flavours are produced at the same rate, according to the standard theory (see (3.24)). Customarily, the cross sections $\sigma(e^+e^- \to X)$ are given in units of $\sigma_{\mu\mu}$, the lowest order QED cross section for $e^+e^- \to \mu^+\mu^-$:

$$\sigma_{\mu\mu} = \frac{4\pi\alpha^2}{3s} \simeq \frac{87}{s} \ [\text{nb}] \qquad (3.44)$$

with s in GeV^2. The cross section, normalised to $\sigma_{\mu\mu}$, is labelled R:

$$R(e^+e^- \to X) = \frac{\sigma(e^+e^- \to X)}{\sigma_{\mu\mu}} . \qquad (3.45)$$

In e^+e^- annihilation one thus can study the production of all three known generations of leptons and test a very central feature of the standard theory, namely universality, which means that the weak neutral current couples to the different lepton families with the same strength (see Chap. 2). In the subsequent chapters the experimental information on the three reactions $e^+e^- \to e^+e^-$ (Bhabha scattering), $e^+e^- \to \mu^+\mu^-$ and $e^+e^- \to \tau^+\tau^-$ will be reviewed. The observables studied are production cross sections, angular asymmetries and, in the case of $\tau^+\tau^-$ production, the polarisation of the final state lepton.

3.4.1 The Reaction $e^+e^- \to e^+e^-$

Although the least sensitive of the leptonic pair production reactions concerning electroweak effects, Bhabha scattering is of great practical importance for all e^+e^- experiments. Bhabha scattering events are the most abundant source for many calibration purposes. Their high rate in the very forward directions enable, e.g., statistically significant measurements of the instantaneous luminosity. Wide angle scattering events have a typical cross section, within the acceptance of the central detector systems, of about $30 - 40$ units of R and are still produced copiously in comparison to multihadronic final states. Bhabha scattering generally used to determine the integrated luminosity and thus serves as the normalisation point for each physics reaction. Since the amplitude for Bhabha scattering is dominated by pure QED, it is calculable to a high degree of precision. Furthermore the statistics of the events is adequate to study many properties of the detector systems used and, most importantly, to derive the energy calibration for the electromagnetic calorimeters.

All PEP/PETRA experiments have measured Bhabha scattering at wide angles (typically $|\cos\theta| \leq 0.85$). High statistics data at 29 GeV have been obtained from HRS (Derrick et al., 1985), MAC (Delfino, 1985), and MARK II (Levi et al., 1983) at PEP. At PETRA, measurements have been performed mainly in the medium energy range $\sqrt{s} \sim 35$ GeV by CELLO (Behrend et al.,

1983), JADE (Bartel et al., 1985), MARK J (Böhm, 1983), PLUTO (Berger et al., 1985), and TASSO (Althoff et al., 1984), as well as in the high energy region around $\sqrt{s} = 44$ GeV by CELLO (Behrend et al., 1985), JADE (Bartel et al., 1985), and TASSO (Althoff et al., 1985). Preliminary results have recently become available from TRISTAN (Unno, 1988).

It would lead too far to give here an account of the analysis procedures used by the individual experiments. Still, a few general remarks seem useful to appreciate the systematic uncertainties involved. In lowest order QED the topology for Bhabha scattering events is very simple, i.e. collinear e^+e^- pairs, each particle having beam energy (E_B). However, initial and final state radiation, which is mostly undetected, introduces an acollinearity between the outgoing leptons and reduces their energy. Furthermore, bremsstrahlung and e^+e^- conversions of the radiated photons in the material surrounding the interaction point may occur and raise the number of particles in the final state. The experiments typically require an acollinearity angle ξ less than $10° - 15°$ and set limits for the number of charged particles in the final state (≤ 6). With respect to the analysis of Bhabha scattering, the experiments can basically be divided into three classes with different sensitivity to background from other reactions such as $e^+e^- \rightarrow \mu^+\mu^-, \tau^+\tau^-$, hadrons. These background sources are particularly harmful in the backward hemisphere, where the cross section is comparable in size to the other leptonic reactions and where electroweak effects are expected to show up. The three classes are:

a) Electron and positron identification using momentum information from curvature measurements in a tracking device as well as energy and position information from an electromagnetic calorimeter (HRS, MARK II, CELLO, JADE, PLUTO). Backgrounds from other processes such as $e^+e^- \rightarrow \tau^+\tau^-$ with both τ's decaying to electrons are typically at a level of 0.5%.

b) Identification of the charge of the final state particles with no further identification of the particles (TASSO). The background due to $e^+e^- \rightarrow \mu^+\mu^-$ is dominant ($\sim 5.5\%$, averaged over all scattering angles) and is subtracted statistically from the angular distribution. In the backward hemisphere this background is substantial.

c) Identification of electrons and positrons as charged showering particles with substantial energy deposition (typically $\geq E_B/3$) in a calorimeter, but without charge determination (MAC, MARK J). The main deficiency of these analyses is the folding of the angular distribution about $\theta = 90°$, thus losing the information about the backward hemisphere. Moreover somewhat higher backgrounds from leptonic and hadronic reactions ($\sim 1\%$) are encountered.

It has become customary to correct for radiative effects besides the usual acceptance and inefficiency corrections so that the resulting differential cross section can be directly compared to the lowest order QED cross section (see (3.40)). For the radiative corrections, which depend strongly on the actual cuts applied in the experiment (see the discussion of Sect. 3.2), Monte Carlo programs following the $O(\alpha^3)$ calculations of Berends et al. (1973, 1976) and Berends and

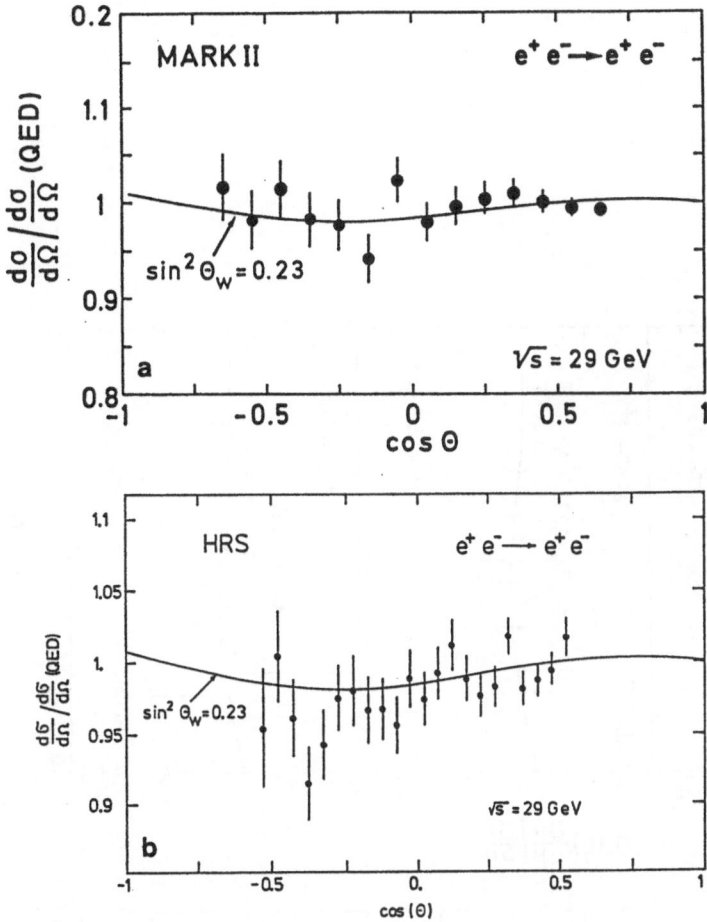

Fig. 3.18. Bhabha cross section relative to lowest order QED from MARK II (a) and HRS
(b). The full curves show the prediction of the standard theory for $\sin^2 \theta_W = 0.23$

Kleiss (1983) are used. Some examples of the fully corrected differential Bhabha
cross sections from the various PEP/PETRA experiments, divided by the lowest
order QED expectation, are shown in Figs. 3.18–20. MARK J (due to lack
of magnetic field in the central detector) and MAC (insufficient bending power
for high momentum showering particles) cannot separate forward and backward
hemispheres.

As demonstrated in Fig. 3.9, all values of $\sin^2 \theta_W$ lead to very small electroweak contributions to the pure QED Bhabha cross section. Different values
for $\sin^2 \theta_W$ only vary slightly in their $\cos \theta$ dependence. In fact, all measurements are compatible with QED at present energies and statistics. Several experiments have analyzed the Bhabha differential cross section in the framework
of the GSW model (using (3.38)) and have obtained typical values for the Weinberg angle in the range $0.12 < \sin^2 \theta_W < 0.38$ (e.g. CELLO: Behrend et al.,

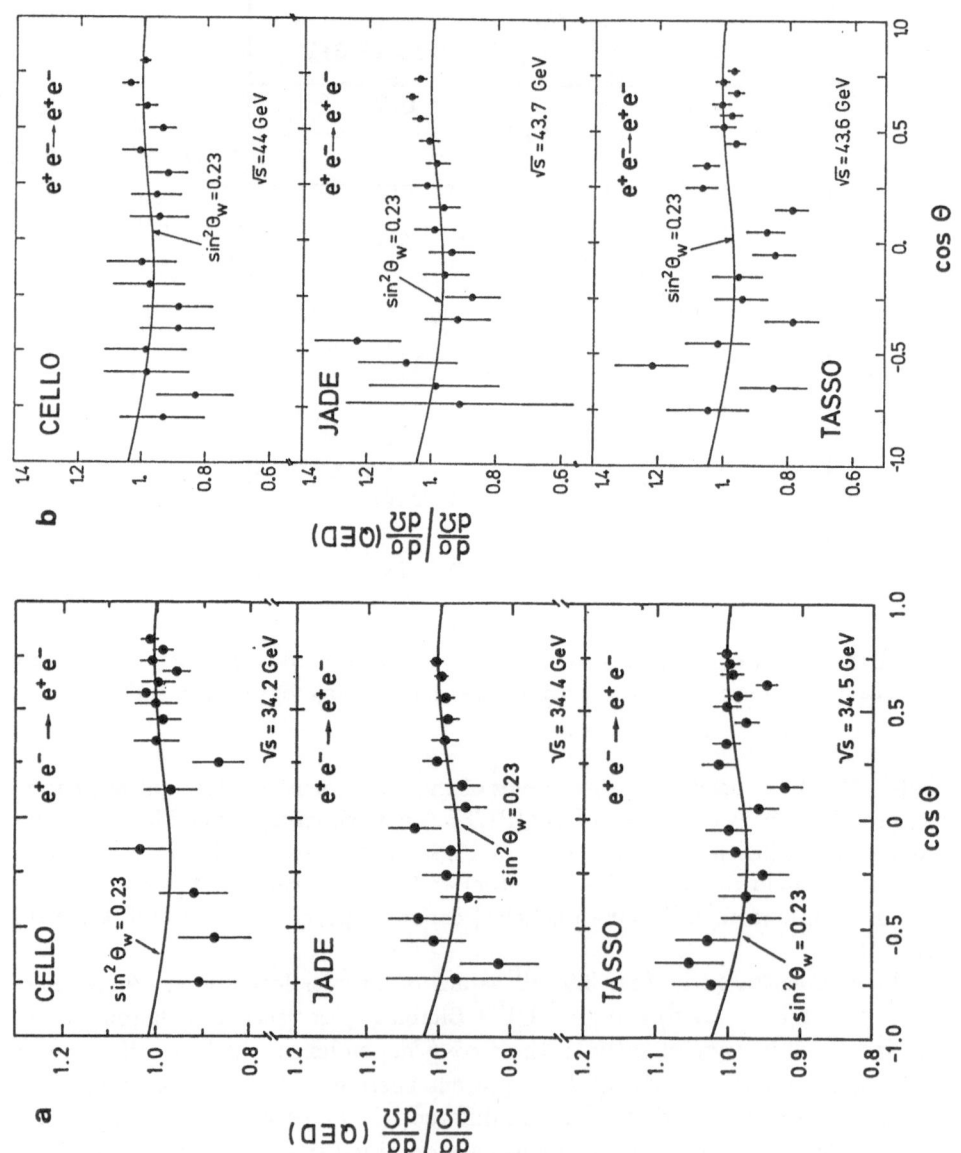

Fig. 3.19. Bhabha cross section relative to lowest order QED from CELLO, JADE, and TASSO. The full curves show the prediction of the standard theory for $\sin^2\theta_w = 0.23$

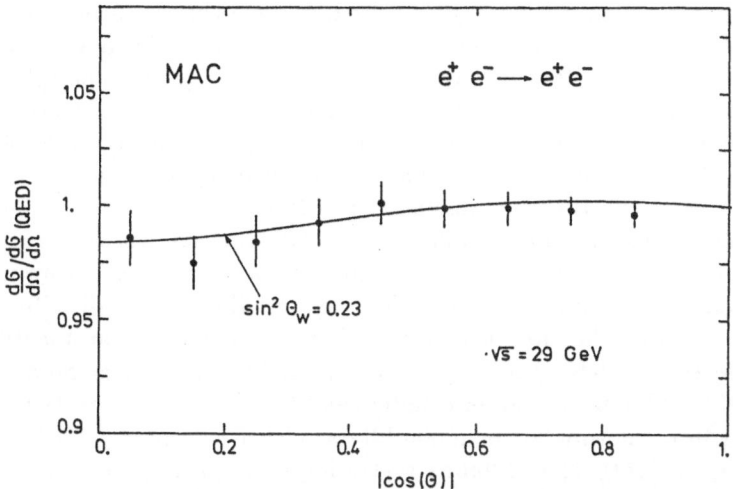

Fig. 3.20. Bhabha cross section relative to lowest order QED from the MAC detector. The full curve shows the prediction of the standard theory for $\sin^2 \theta_W = 0.23$

1981a; JADE: Bartel et al., 1983). TASSO (Althoff et al., 1985) determine a value of $\sin^2 \theta_W = 0.23 \pm 0.08$.

Although inclusion of the neutral current exchange to the Bhabha scattering amplitude improves the fit to the experimental data, no significant weak effect can be established from the Bhabha scattering process alone. On the other hand, deviations from point-like QED have traditionally been parametrised by form factors $F(s)$, $F(t)$, multiplying the s and t channel amplitudes in Table 3.2,

$$F(s) = 1 \pm \frac{s}{s - \Lambda_{\pm}^2}, \quad F(t) = 1 \pm \frac{t}{t - \Lambda_{\pm}^2}.$$

Here, $\Lambda_{\pm}$ are cut-off parameters (Drell, 1965), indicative of an energy scale, where deviations from the pointlike nature of the electromagnetic coupling could become apparent. Evolutions of the QED cut-off parameters from the highest PETRA energies, including statistical and systematic errors, indicate typical values for $\Lambda_{\pm}$ of order 200 GeV (Kiesling, 1985). Further analyses to test the standard theory using the data on Bhabha scattering will be presented in subsequent sections.

3.4.2 The Reaction $e^+e^- \rightarrow \mu^+\mu^-$

The production of muon pairs yielded early indications for the presence of weak neutral currents in e^+e^- annihilation. In 1982, the PETRA experiments (Adeva et al., 1982; Behrend et al., 1982; Bartel et al., 1982; Brandelik et al., 1982) reported evidence for deviations from QED in the angular distribution of the outgoing muons with respect to the beam direction. Measurements of this charge asymmetry, still limited in their statistical significance, were in full accord with the standard theory. Since then PEP and PETRA experiments have continuously increased their data samples so that the individual experiments by now have etablished a better than 4 standard deviation effect in the charge asymmetry.

Muon pair final states are extracted from the experimental data by selecting events with collinear ($\xi < 10° - 15°$), high momentum ($p > E_B/3$) pairs of charged particles. The reaction $e^+e^- \to \mu^+\mu^-$ is topologically very similar to Bhabha scattering. Since the Bhabha cross section is larger by a factor of ~ 30 compared to the $\mu^+\mu^-$ cross section within $|\cos\theta| < 0.85$ and has a strong positive charge asymmetry (due to the t-channel exchange), a contamination from even a small fraction of Bhabha events will severely distort the angular distribution and has to be removed therefore with high efficiency. In some experiments (CELLO, JADE, MAC, MARK J) muons are positively identified by small energy depositions (≤ 2 GeV) in the shower counters and by requiring the particles to penetrate the hadron calorimeters or muon filters. Other experiments (HRS, MARK II, PLUTO, TASSO) reject Bhabha scattering by requiring a calorimeter response for each charged particle consistent with minimum ionisation. This requirement is, in general, sufficient to isolate μ pairs, because pair production of other minimum ionising particles (pions) is found to be negligible, and fluctuations in the shower of a high energy electron to mimic a small energy deposition are rare. Typically less than 0.5% of the final muon sample consist of Bhabhas having passed the above rejection criteria.

Another important source of background comes from cosmic muons crossing the interaction region, thus simulating a $\mu^+\mu^-$ pair. Most experiments are equipped with scintillation counters surrounding the central detector for event time and time-of-flight (TOF) measurements to reject cosmic rays. The CELLO detector, due to space limitations between coil and calorimeter, does not have a TOF counter system. Cosmic ray background is removed here by statistical subtraction, extrapolating experimental distributions of cosmic ray events which miss the interaction region. Other reactions, such as $e^+e^- \to e^+e^-\mu^+\mu^-$ with both final state electrons undetected, or $e^+e^- \to \tau^+\tau^-$ with both τ's decaying into muons and neutrinos may contaminate the muon sample. Restriction to high momentum particles, coplanar with the incoming beam, reduces these backgrounds to levels around and below 1 percent. The residual background which cannot be removed by experimental cuts is corrected for, usually by Monte-Carlo methods.

Cross Section

Following the discussion in Sect. 3.1.2 on e^+e^- annihilation into lepton-antilepton pairs the weak neutral current contributions to the total cross section expected in the standard theory are exceedingly small, even at the highest PETRA energies. These contributions are from electroweak interference which is proportional to the (small) vector charges v_i, and from the purely weak term, proportional to the sum of the squared weak charges $v_i{}^2$ and $a_i{}^2$, but damped by the Z^0 propagator. Expressed in units of the QED cross section (3.44) the cross section (3.24) for $e^+e^- \to \mu^+\mu^-$ as a function of the squared centre of

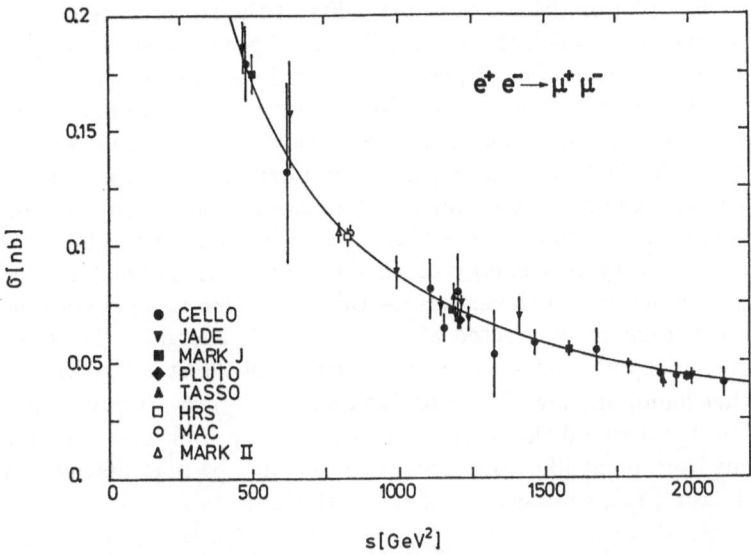

Fig. 3.21. Measurements of the total cross section for the reaction $e^+e^- \rightarrow \mu^+\mu^-$ as a function for the square of the centre of mass energy. The lowest order QED prediction is drawn as a solid line

mass energy s reads:

$$R_{\mu\mu}(s) = 1 + v_e v_\mu \operatorname{Re}\{g(s)\} + \frac{1}{4}(v_e^2 + a_e^2)(v_\mu^2 + a_\mu^2)|g(s)|^2. \qquad (3.46)$$

Note that the electroweak interference term is negative below the Z^0, due to the sign of $\operatorname{Re}\{g(s)\}$. With $\sin^2\theta_W$ close to 1/4, the effect expected at PEP/PETRA energies is around the percent level. Even at PETRA's top end of about 46 GeV the contributions from electroweak interference and the purely weak term are only -0.15% and $+1.4\%$, respectively, where the size of the positive purely weak contribution always dominates the one from interference. Measurements on $\sigma_{\mu\mu}$ from both PEP and PETRA are shown in Fig. 3.21. Data are available from CELLO (Behrend et al., 1982), JADE (Bartel et al., 1982, 1985), MAC (Fernandez et al., 1983; Ash et al., 1985), MARK II (Levi et al., 1983), MARK J (Adeva et al., 1983, 1985), PLUTO (Berger et al., 1983) and TASSO (Althoff et al., 1984). The individual experiments have corrected their data for radiative effects. The errors shown include the statistical uncertainties only. Within these errors, $\sigma_{\mu\mu}$ is in perfect agreement with QED. Systematic errors, dominated by the luminosity measurement, are typically of order $2 - 4\%$. It is clear that the size of these systematic errors precludes a test of the standard theory at PEP/PETRA, and even at TRISTAN energies. The corrections to the total cross section of leptonic final states due to the Z^0 exchange are well below the systematic uncertainties up to centre of mass energies of order 50 to 60 GeV.

With $\sin^2\theta_W = 0.23$ the standard theory predicts $v_{e,\mu} = -0.08$, small compared to the axial charges, which are predicted to be $a_{e,\mu} = -1$ (see Chap. 2).

We will see lateron how the prediction of universality of the leptonic weak couplings, i.e. the equality of the weak charges for the three generations of charged leptons, can be tested for the axial couplings with the help of the angular charge asymmetries. The total cross section, in principle, could offer a complementary test for the vector couplings. Assuming the standard values for the weak charges of the electron current in the initial state and the standard axial charge for the final state muon current, the muonic vector coupling can be limited only within the interval $-4 \leq v_\mu \leq 1.5$, when an uncertainty of 2 % is assumed for the cross section measurement at a typical energy of 35 GeV. Due to the smallness of the weak effects in comparison to the experimental errors even large deviations from universality will remain undetected at PETRA/PEP energies. Thus the total cross section cannot be used as a test of this aspect of the standard theory either. On the other hand, any significant deviation from $R_{\mu\mu} = 1$ would signal deviations from QED and would therefore point towards new and unanticipated physics. Deviations from point-like QED cross sections may be parametrised by means of a form factor $F(s)$, similar to Bhabha scattering,

$$F(s) = 1 \pm \frac{s}{s - \Lambda_\mp^2} \ . \tag{3.47}$$

The mass scale $\Lambda_\mp$ can be viewed as the inverse of a length below which the QED coupling to fermions can no longer be considered point-like. Lower limits for $\Lambda_\mp$ derived from $R_{\mu\mu}$, including systematic errors, are of order 200 GeV for the PETRA experiments (see, e.g., Kiesling, 1985) which means that muons interact point-like, i.e. do not exhibit structure down to distances of order 10^{-16} cm.

Charge Asymmetry

As shown in Sect. 3.1, weak neutral currents introduce terms linear in $\cos\theta$ in the differential cross section for $e^+e^- \rightarrow \mu^+\mu^-$. One can thus derive an angular or charge asymmetry $A_{\mu\mu}$ which is proportional to the product of the axial vector coupling constants $a_e a_\mu$ (compare (3.20)):

$$A_{\mu\mu} = \frac{3}{4} \, a_e a_\mu \, \mathrm{Re}\{g(s)\} + \text{purely weak terms} \ . \tag{3.48}$$

In contrast to the total cross section, the purely weak terms in the expression for the charge asymmetry can be safely neglected at PEP/PETRA energies. Since $a_e a_\mu = 1 \gg v_e v_\mu$ a measurable effect in $A_{\mu\mu}$ should show up, unlike the case of $R_{\mu\mu}$ discussed above. For the PEP centre of mass energy of 29 GeV the standard theory predicts, on the Born level, a charge asymmetry of -5.8%, while at PETRA for $\sqrt{s} = 35$ GeV and 44 GeV values of -8.9% and -15.6%, respectively, are expected. At TRISTAN ($\sqrt{s} = 54$ GeV), the charge asymmetry expected is as large as -26.8%.

For the measurement of the charge asymmetry the understanding of systematic effects is of central importance. The major contaminating reactions have already been discussed in the previous section. In addition, one has to make sure that the charges of the outgoing muons are determined correctly, apart from the problem of a possible charge-dependent acceptance asymmetry, as discussed in Sect. 3.3. Possible biases in the determination of the muon charge are related to the finite momentum resolution for charged particles in the tracking device: Due to measurement errors in the reconstructed space points, the sign of the curvature and thus the charge sign of the muon can be falsely determined. Note that additional hard photons in the final state from radiative effects will lower the momenta of the muons. Like-sign muon pair events (one muon charge identified incorrectly) can usually be classified according to the charge of the lower-momentum muon, which has a better chance to be measured correctly. When *both* muons receive the wrong charge sign the resulting angular charge asymmetry will be "washed out", i.e. reduced in magnitude. The fraction of wrong charge sign determinations, as determined from the like-sign muon pair events, is usually below 1.0%. Therfore the fraction of events where the wrong charge has been assigned to both muons is negligible in most experiments.

While these purely experimental biases are found to be small, a sizeable charge asymmetry of about +1.5% is expected from radiative corrections (see Sect. 3.2). These corrections are taken into account when the electroweak charge asymmetry is extracted from the data. All experiments, except MARK II, have chosen to present radiatively corrected differential cross sections, using either the $O(\alpha^3)$ QED corrections to the one-photon-exchange diagram (Berends and Kleiss, 1981) or the "full" QED correction including also radiative corrections to the Z^0 exchange diagram (Berends et al., 1982). The differential cross sections for the reaction $e^+e^- \rightarrow \mu^+\mu^-$ for the PEP experiments at $\sqrt{s} = 29$ GeV are shown in Fig. 3.22. The MAC collaboration (Ash et al., 1985) has accumulated the largest statistics, amounting to about 16 000 events from 222 pb^{-1}, whereas HRS (Baranko et al., 1984; Derrik et al., 1985) have analysed 106 pb^{-1} so far. The data from MARK II (Levi et al., 1983) are based on 100 pb^{-1}. While MAC chose the correct their data for full QED, HRS used reduced QED corrections and MARK II did not apply any correction for QED at all. Therefore the QED expectations for MARK II in Fig. 3.22, e.g., does not correspond to the lowest order, but rather to $O(\alpha^3)$.

Figures 3.23,24 show the PETRA data taken at $\sqrt{s} = 34.5$ GeV by JADE (Bartel et al., 1985), MARK J (Adeva et al., 1985), PLUTO (Berger et al., 1985), and TASSO (Althoff et al., 1984). High energy data, centering around $\sqrt{s} = 44$ GeV have been presented by CELLO (Behrend et al., 1985), JADE (Bartel et al., 1985), MARK J (Adeva et al., 1985), and TASSO (Althoff et al., 1985). The full curves drawn in the figures result from fits of the electroweak cross section (3.15) to the data while the dashed curves represent the symmetric lowest order QED expectation.

The angular distributions from the PETRA experiments agree well with each other, as do the PEP experiments. All differential cross sections clearly deviate from the symmetric lowest order QED expectation and exhibit a negative angular asymmetry as predicted by the standard theory. The asymmetry $A_{\mu\mu}$, as calculated from the fitted coefficients describing the differential cross section

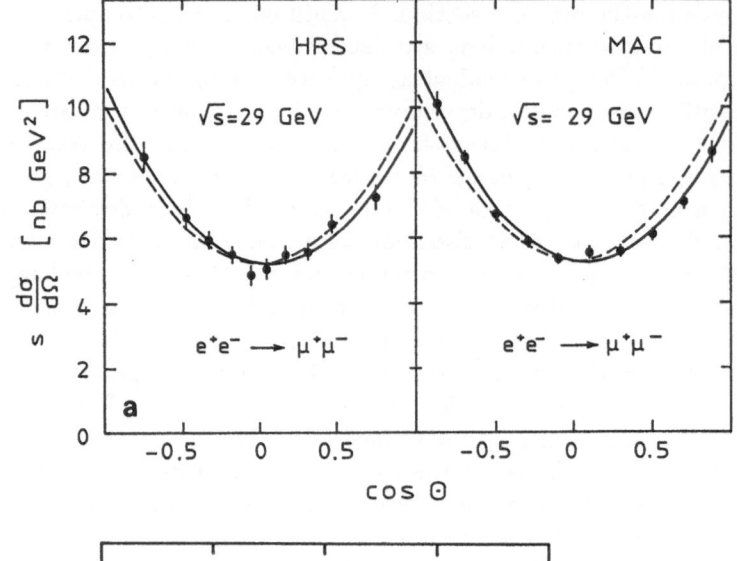

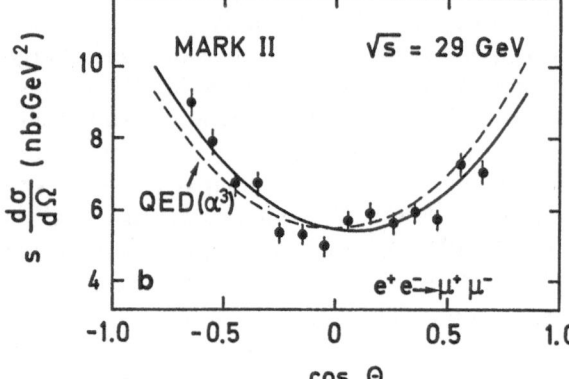

Fig. 3.22. Measurements of the differential cross section for the reaction $e^+e^- \to \mu^+\mu^-$ for the PEP experiments. The dashed curves show the QED expectations (lowest order for MAC and HRS, $O(\alpha^3)$ for MARK II). The full curves result from fits to the data

in lowest order (see (3.15)), are listed in Table 3.4. In contrast to the differential cross sections shown in the figures, the values for the charge asymmetries $A_{\mu\mu}$ are derived from the data applying $O(\alpha^3)$ QED corrections ("reduced QED") only. For most experiments the error on $A_{\mu\mu}$ is separated in a statistical and a systematic part. The latter, however, does generally not include the uncertainty due to higher order weak (loop) corrections (see the discussion in Sect 3.2). Also given in Table 3.4 is the expectation for the asymmetry in the lowest order of the standard theory using a mass for the Z^0 of 92 GeV and $\sin^2\theta_W = 0.23$ in the propagator term (3.9) of the Glashow scheme. First results from TRISTAN at $\sqrt{s} = 54.5$ GeV are given in Table 3.5 (Kamae, 1988). These data are also in very good agreement with the expectations from the standard theory.

Figure 3.25 shows the charge asymmetry measurements from the various experiments as a function of the square of the center-of-mass energy. The line at $A_{\mu\mu} = 0$ corresponds to the lowest order QED expectation. Also shown in the figure is the prediction from the standard theory for $\sin^2\theta_W = 0.23$ and $M_Z = 92$ GeV according to the parametrisation (3.9). The individual experiments are in

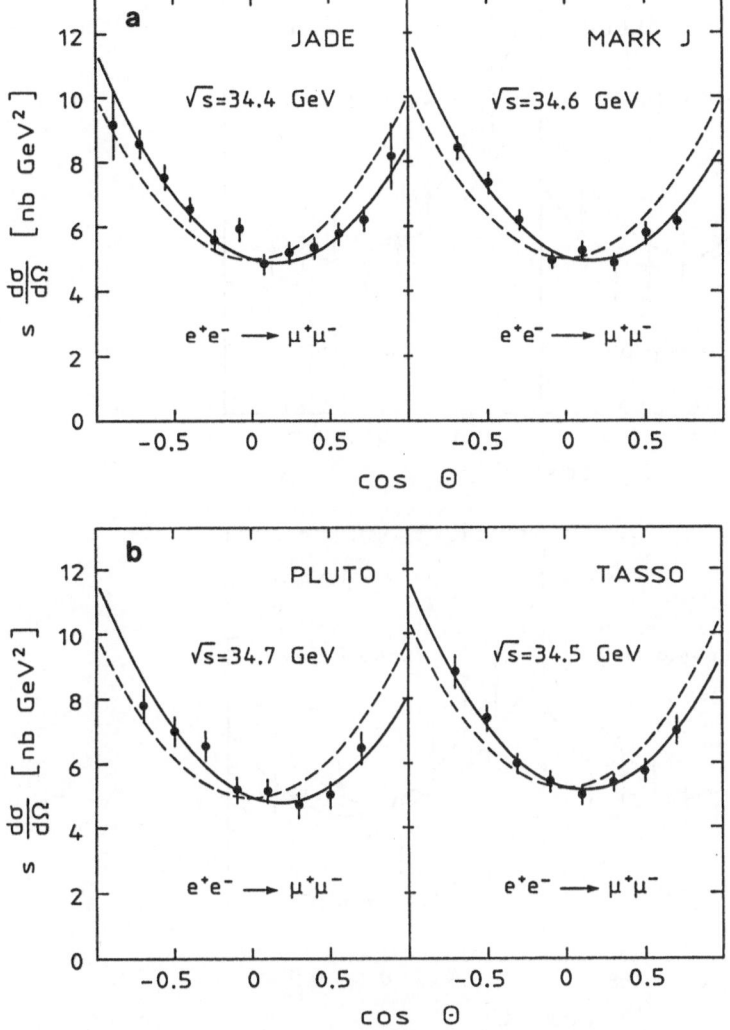

Fig. 3.23. Measurements of the differential cross section for the reaction $e^+e^- \rightarrow \mu^+\mu^-$ for the PETRA experiments at $\sqrt{s} = 34.5$ GeV. The dashed curves show the lowest order QED expectation. The full curves result from fits to the data

agreement with the theoretical prediction, although a slight preference for larger asymmetries is noticeable in the PETRA energy range. On the other hand, the PETRA data from 1986, the last year of running, at $\sqrt{s} = 35$ GeV are in perfect agreement with the expectation (see Table 3.4, 5).

To further investigate the significance of the observed effect, i.e. deviation from the symmetric angular distribution expected from QED and agreement with the predictions of the standard theory, averages have been given in the Tables 3.4,5 for $A_{\mu\mu}$ for the different PEP, PETRA, and TRISTAN energy ranges. These averages have been derived by weighting the individual experiments by

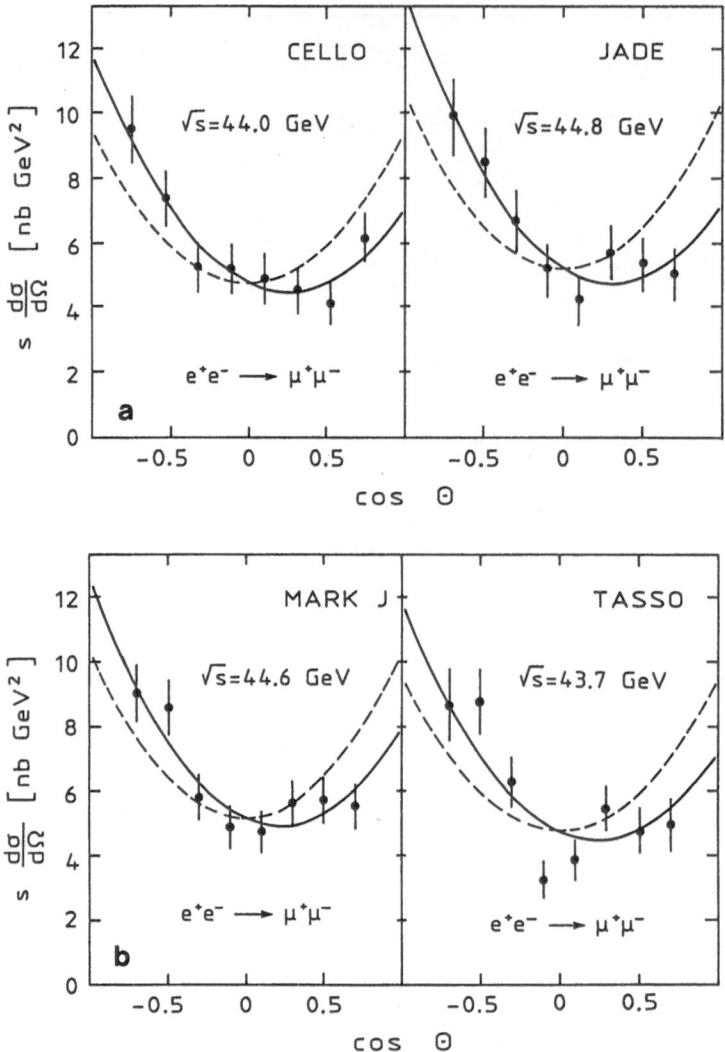

Fig. 3.24. Measurements of the differential cross section for the reaction $e^+e^- \to \mu^+\mu^-$ for the PETRA experiments at $\sqrt{s} = 44.0$ GeV. The dashed curves show the lowest order QED expectation. The full curves result from fits to the data

their statistical errors. Concerning the estimate for the systematical errors of the averaged asymmetries it does not seem justified to consider them as uncorrelated. Many procedures in the extraction of the corrected differential cross sections for μ pair production are very similar or even identical for the individual experiments, such as the use of Monte-Carlo programs for radiative corrections, event generators for the estimation of multihadronic and QED background, various shower simulation programs etc.. In that spirit the systematic error of the PEP/PETRA averages have been estimated of the same order as for the individual experiments. It is clear that this simple procedure is inadequate to estimate

Table 3.4. Measurements of the charge asymmetry for the reaction $e^+e^- \to \mu^+\mu^-$ from PEP and PETRA. The predictions from the standard theory are given in the last column, with $\sin^2\theta_W = 0.23$ and $M_Z = 92$ GeV

Exp.	$\sqrt{s}$ [GeV]	$\int \mathcal{L}\, dt$ [pb^{-1}]	$N_{\mu\mu}$	$\varepsilon_{\mu\mu}{}^*$	$A_{\mu\mu}$ [%]	A_{GSW} [%]
HRS	29	106	5057	0.46	$-\ 4.9 \pm 1.5 \pm 0.5$	-5.8
MAC	29	222	16058	0.70	$-\ 5.7 \pm 0.7 \pm 0.2$	-5.8
MARK II	29	100	5312	0.51	$-\ 7.1 \pm 1.7$	-5.8
PEP	29	428	26427	0.60	$-\ 5.8 \pm 0.6 \pm 0.5$	-5.8
CELLO	34.2	11	387	0.46	$-\ 6.4 \pm 6.4$	-8.5
JADE	34.4	71	3400	0.65	$-11.1 \pm 1.8 \pm 1.0$	-8.6
MARK J	34.6	76	3658	0.66	$-11.7 \pm 1.7 \pm 1.0$	-8.7
PLUTO	34.7	44	1553	0.49	$-13.3 \pm 3.0 \pm 1.0$	-8.7
TASSO	34.5	75	2673	0.49	$-\ 9.1 \pm 2.1 \pm 0.5$	-8.6
PETRA	34.5	277	11671	0.58	$-11.0 \pm 1.0 \pm 1.0$	-8.6
CELLO	35.0	83	3450	0.47	$-\ 8.9 \pm 2.0 \pm 1.0$	-8.9
JADE	35.0	82	3901	0.67	$-10.9 \pm 1.7 \pm 1.0$	-8.9
MARK J	35.0	68.0	3196	0.66	$-\ 8.1 \pm 1.9 \pm 0.5$	-8.9
TASSO	35.0	108.5	2563	0.33	$-10.6^{+2.2}_{-2.3} \pm 0.5$	-8.9
PETRA	35.0	341.5	13110	0.54	$-\ 9.6 \pm 1.0 \pm 1.0$	-8.9
CELLO	39.0	12.4	288	0.41	$-\ 4.8 \pm 6.5 \pm 1.0$	-11.5
JADE	38.0	11.9	422	0.59	$-\ 9.7 \pm 5.0 \pm 1.0$	-10.8
MARK J	39.2	16.9	671	0.70	$-10.6 \pm 4.0 \pm 0.5$	-11.7
TASSO	38.3	8.8	173	0.33	$+\ 1.7^{+8.5}_{-8.6} \pm 0.5$	-11.0
PETRA	38.7	50	1554	0.54	$-\ 8.1 \pm 2.7 \pm 1.0$	-11.3
CELLO	44.0	20.5	611	0.66	$-18.8 \pm 4.5 \pm 1.0$	-15.5
JADE	43.7	35	1258	0.79	$-19.1 \pm 2.8 \pm 1.0$	-14.7
MARK J	44.1	43.4	1278	0.66	$-15.8 \pm 2.8 \pm 0.5$	-15.1
TASSO	43.6	35.2	612	0.38	$-17.6^{+4.4}_{-4.3} \pm 0.5$	-15.2
PETRA	43.9	134	3759	0.62	$-17.6 \pm 1.7 \pm 1.0$	-15.4

*   overall efficiency for muons assuming $R_{\mu\mu} = 1$.

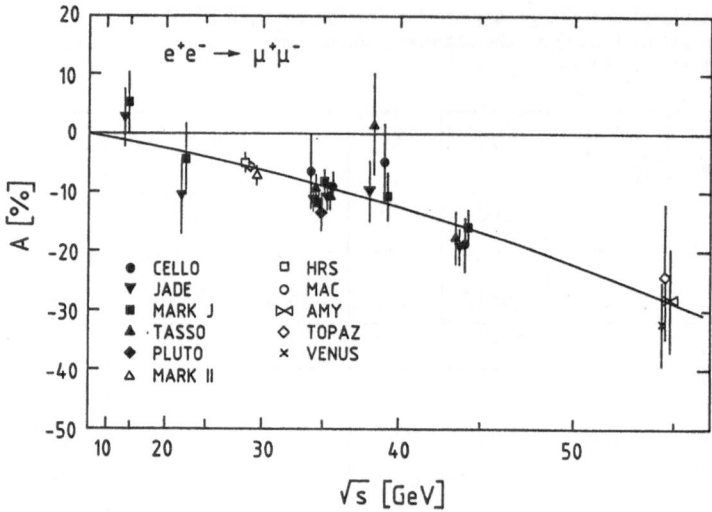

Fig. 3.25. Charge asymmetry measurements for the reaction $e^+e^- \to \mu^+\mu^-$ as function of $s = E_{\text{CM}}^2$. The curve corresponds to the lowest order expectation from the standard theory with $M_Z = 92$ GeV and $\sin^2\theta_W = 0.23$

the true error of the PEP/PETRA averages but may be considered at least as some improvement over the naive procedure of adding both statistical *and* systematic inverse errors in quadrature. Looking at the averages and their errors thus obtained one finds perfect agreement of the measurements at PEP, PETRA and TRISTAN with the expectations of the standard theory. The interpretation of the observed asymmetries in terms of the weak charges and the other parameters in the standard theory will be taken up after discussing the production of leptons of the third generation, i.e. the reaction $e^+e^- \to \tau^+\tau^-$.

3.4.3 The Reaction $e^+e^- \to \tau^+\tau^-$

Since its discovery by Perl et al. (1975) the τ lepton has been investigated for properties discriminating it from the electron or the muon. Except for phenomena related to its higher mass ($m_\tau = 1.784\,\text{GeV}/c^2$) no difference has yet been found. In the framework of the standard theory the τ is accomodated for by postulating a third doublet of leptons (ν_τ, τ) in addition to the electron and the muon doublets (see Chap. 2). This interpretation of the τ as a new "sequential" lepton has immediate consequences for its properties. In particular, the lifetime and the decays should be determined by the charged current, and the interactions of the τ with the neutral current should be identical to those of the electron and muon. By measuring the total cross section and the charge asymmetry of the reaction $e^+e^- \to \tau^+\tau^-$ one may thus test the universal coupling of the neutral weak current to the leptons. There is, however, an additional merit in studying τ pair production: the weak decay of the τ can be used to analyze its potential longitudinal polarisation, which provides the only handle at present to measure the vector coupling constant of the τ to the weak neutral current.

Table 3.5. New measurements of the charge asymmetry for the reaction $e^+e^- \rightarrow \mu^+\mu^-$ from TRISTAN. The predictions from the standard theory are given in the last column, with $\sin^2\theta_W = 0.23$ and $M_Z = 92$ GeV

Exp.	$\sqrt{s}$ [GeV]	$A_{\mu\mu}$ [%]	A_{GWS} [%]
AMY	54.5	-28.1 ± 8.7	-27.5
TOPAZ	54.5	-24.1 ± 10.4	-27.5
VENUS	54.3	-32.0 ± 7.0	-27.3
TRISTAN	54.5	-29.1 ± 4.8	-27.5

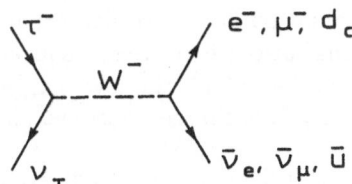

Fig. 3.26. First order diagrams for τ decay in the standard theory

Like the muon, the τ is assumed to decay only through the weak charged current, coupling to its own neutrino ν_τ which implies conservation of the τ lepton quantum number. Figure 3.26 shows the diagram for the τ decays in the standard theory. Purely leptonic decays ($\tau^- \rightarrow e^- \bar{\nu}_e\nu_\tau$, $\mu^- \bar{\nu}_\mu\nu_\tau$) and semihadronic decays ($\tau^- \rightarrow d_c \bar{u}\nu_\tau$) are expected. Note that only quarks from the first generation are allowed energetically. From these diagrams one can get a crude estimate of the purely leptonic and semihadronic branching ratios by observing that the quark diagram occurs in three different colours, amounting to five possible decay channels of equal probability when mass effects and hadronic final state interactions are neglected. One thus expects the purely leptonic decays at a level of 20% each, and the sum of all semihadronic decays at $\sim 60\%$. The purely leptonic decay widths can be calculated precisely in the framework of the standard theory. Semihadronic decay widths $\Gamma_i(\tau \rightarrow \text{hadrons} + \nu_\tau)$ have been calculated by a number of authors (Thacker and Sakurai, 1971; Tsai, 1971; Gilman and Miller, 1978; Kawamoto and Sanda, 1978; Pham et al., 1978; Tsai, 1980; Gilman and Rhie, 1985) on the basis of the standard theory, using additional theoretical and experimental input. One such input is, e,g,, the conserved vector current (CVC) hypothesis, which relates the vector part of the W current to the isovector part of the electromagnetic current. It is interesting to note that most of the above cited calculations were done long before the τ was discovered.

There is good theoretical agreement on the partial widths for the purely leptonic decays and semihadronic two-body decays such as $\tau \rightarrow \pi\nu$, $K\nu$, which can be calculated from "first principles". The decays into an even number of pions can be derived, invoking CVC, from the corresponding $e^+e^- \rightarrow 2n\pi$ cross sections, suitably corrected to include only the isovector part. The axial vector current leads to hadronic final states with an odd number of pions where less certain theoretical concepts, such as the partially conserved axial-vector current (PCAC) hypothesis, sum rules etc., are employed in the calculation of $\Gamma_i(\tau \rightarrow$ hadrons $+ \nu_\tau)$. Since the branching ratio $B_i(\tau \rightarrow X_i\nu_\tau)$ is defined as

$$B_i(\tau \rightarrow X_i\nu) = \frac{\Gamma_i(\tau \rightarrow X_i\nu)}{\Gamma_{\text{tot}}} , \qquad (3.49)$$

the uncertainties in the decays mediated by the axial vector current will also affect the predictions for the branching ratios of the well-understood decays such as $\tau \rightarrow e\bar{\nu}\nu$, $\mu\bar{\nu}\nu$.

Despite these uncertainties there is a clear prediction that the τ will decay predominantly into one charged particle plus neutrals. This so-called one-prong topology is expected to have a branching ratio around 85 %. Since five-prong decays are rare (see below), the three-prong branching ratio is expected at a level of about 15 %. In selecting events of the reaction $e^+e^- \rightarrow \tau^+\tau^-$ the low charge multiplicity (about 75 % have two, the rest has mainly four charged particles) is a distinctive feature of τ pair production.

Experimentally, the process $e^+e^- \rightarrow \tau^+\tau^-$ is considerably more complicated than the two leptonic reactions discussed in the preceeding chapters. Due to its short lifetime ($T_\tau \sim 3 \times 10^{-13}$ s) the τ decays before detection and a quantitative understanding of the branching ratios for the various decays is necessary to be able to measure cross sections and observables derived from them. It thus seems appropriate to deviate from the general discussion of neutral weak currents and mention recent measurements concerning the properties of the τ lepton related to the weak charged current. These measured properties verify the sequential nature of the τ lepton as postulated by the standard theory.

A considerable amount of information about τ decays has been obtained already at SPEAR (Bacino et al., 1979; Jaros et al., 1978; Dorfan et al., 1979, 1981; Blocker et al., 1982a) and DORIS (Alexander et al., 1978; Brandelik et al., 1978; Bartel et al., 1978). For a review of the τ properties known from the SPEAR and DORIS measurements see, e.g., Perl (1980) and Kalmus (1982). Since then the new generation of e^+e^- machines and the considerably improved statistics have significantly increased the knowledge on the τ lepton. In particular in the PEP/PETRA energy range the τ pair signature becomes more distinct from most background reactions for a number of reasons:

— The τ lepton decays into final states containing predominantly one or three charged particles. The total charged multiplicity of τ events is therefore small compared to the average charged multiplicity of hadronic events at and beyond the PEP/PETRA energies.
— Due to the higher laboratory momenta of the τ leptons their decay products maintain the original direction of flight of the parent τ's within a narrow angular cone, leading to almost back-to-back event topologies.

— The τ decay products, as a consequence of the low multiplicity, generally have large momenta and may thus be separated from two photon reactions which lead to events with low multiplicity of low momentum particles.

A severe background, however, arises from higher order QED reactions such as $e^+e^- \to e^+e^-\gamma(\gamma)$, which has a number of unpleasant features. First of all, the cross section for these events is large compared to $e^+e^- \to \tau^+\tau^-$. Secondly, the topology of the charged particles fully simulates τ pair events: the electron tracks are nearly back-to-back, with moderate momenta. Furthermore a photon may convert to an e^+e^- pair in the material in front of the tracking chambers and simulate a 3-prong final state, thus influencing the proper measurement of the topological branching fractions. Most important for the study of electroweak effects is the fact that the angular distribution is very similar to that of non-radiative Bhabha scattering, with a very strong positive forward-backward asymmetry caused by the t-channel photon exchange (see Sect 3.4.1).

In order to isolate the reaction $e^+e^- \to \tau^+\tau^-$ from the competing backgrounds, the individual experiments employ different strategies depending on their ability to identify leptons and hadrons:

a) Experiments with good electromagnetic calorimetry and momentum analysis of charged particles accept all τ pair topologies. Higher order QED background ($e^+e^- \to e^+e^-\gamma$, $\mu^+\mu^-\gamma$) is removed event by event via identification of electrons and minimum ionising particles. As a consequence usually those τ pair events are rejected where both τ leptons decay into electrons or muons, giving up on $\sim 6\%$ of the total cross section. The hadronic background can be removed by topological and kinematical cuts, where the most important ones are the total multiplicity of charged particles in the final state (limited, e.g., to ≤ 6) and the invariant masses of the individual back-to-back jets which should not, within the experimental resolution, exceed the mass of the τ.

b) Experiments with good charged particle detection only, restrict themselves in most cases to the topolgy where one τ decays into one charged + neutral particles and the other into three charged + neutral particles (1–3 topology). Background from higher order QED processes is strongly reduced. Contamination from hadronic events becomes more important. Due to the restriction to the 1–3 topology only about 25% of all τ decays are retained. Correspondingly, for cross section measurements one has to rely on the knowledge of the topological branching ratios and is thus very sensitive to uncertainties in these ratios (see below).

c) Experiments with good hadron calorimetry and muon filters select events where one τ is decaying into $\mu\bar{\nu}\nu_\tau$ and the other produces electromagnetic or hadronic showers. Compared to (b), background from hadronic events is suppressed due to the additional muon requirement. The visible cross section is of order 25% of the total cross section, but depends on details of the calorimeters since showering particles are required. Again, the cross section measurement requires the knowledge of the branching ratio for $\tau \to \mu\bar{\nu}\nu_\tau$.

Table 3.6. Recent measurements at PEP/PETRA of the topological branching ratios of the τ lepton in percent

Experiment	Br (1 prong)	Br (3 prong)	Br (5 prong)
CELLO	$84.9 \pm 0.4 \pm 0.3$	$15.0 \pm 0.4 \pm 0.3$	$0.16 \pm 0.13 \pm 0.04$
JADE	$86.1 \pm 0.5 \pm 0.9$	$13.6 \pm 0.5 \pm 0.8$	$0.3 \pm 0.1 \pm 0.2$
PLUTO	$87.8 \pm 1.3 \pm 3.9$	$12.2 \pm 1.3 \pm 3.9$	
TASSO	$84.7 \pm 1.1^{+1.6}_{-1.3}$	$15.3 \pm 1.1^{+1.3}_{-1.6}$	
DELCO	$87.9 \pm 0.5 \pm 1.2$	$12.1 \pm 0.5 \pm 1.2$	
HRS	$86.9 \pm 0.2 \pm 0.3$	$13.0 \pm 0.2 \pm 0.3$	0.13 ± 0.04
MAC	$86.7 \pm 0.3 \pm 0.6$	$13.3 \pm 0.3 \pm 0.6$	$< 0.17(95\% \text{ C.L.})$
MARK II	$87.2 \pm 0.5 \pm 0.8$	$12.8 \pm 0.5 \pm 0.8$	$0.16 \pm 0.08 \pm 0.04$
TPC	84.7 ± 1.0	15.1 ± 1.0	$< 0.3 \ (90\% \text{ C.L.})$

Branching Ratios

The nearly complete detection of τ final states by some PEP/PETRA experiments led to considerable improvement of the knowledge on branching ratios over older results from SPEAR and DORIS (see Particle Data Group, 1982). CELLO (Behrend et al., 1982a) was the first experiment to present new values for $\tau \to$ one-prong + neutrals and $\tau \to$ three-prongs + neutrals, which were hardly in agreement with previous data. The branching ratio $\tau \to$ one-prong + neutrals has increased to $\sim 85\%$ (from about 65 %) while $\tau \to$ three-prongs + neutrals correspondingly decreased to $\sim 15\%$. Subsequently MARK II (Blocker et al., 1982a) and TPC (Aihara et al., 1984) have confirmed these values. Recent precise measurements by MAC (Fernandez et al., 1985), HRS (Akerlof et al., 1985) yield values close to 87% for the one-prong and 13% for the three-prong topology with statistical errors around 0.3%. HRS (Beltrami et al., 1985) was also the first experiment to measure the branching ratio into five charged particles. New precise values for the topological branching ratios have been determined by CELLO. Table 3.6 lists the recent measurements for the topological branching ratios from the cited and various other experiments (JADE: Bartel et al., 1985b; PLUTO: Berger et al., 1985a; TASSO: Althoff et al., 1985a; DELCO: Ruckstuhl et al., 1986; MARK II: Burchat et al., 1985; Schmidtke et al., 1986; TPC: Aihara et al., 1986). It should be noted that the product of the branching ratios for one-prongs and three-prongs, which is used for the total cross section measurement by some experiments, is about $30 - 40\%$ lower compared to the 1982

Table 3.7. Recent measurements of exclusive τ branching ratios from PEP and PETRA in percent. The "World av." do not include older SPEAR/DORIS measurements. The predictions are based on the standard interpretation of the τ as a sequential lepton and are normalised to the electronic branching ratio. Mesons in semihadronic decays are charged unless explicitly stated otherwise

Decay channel	Br	Experiment	Prediction
$\tau \to e\,\nu\,\nu$	18.1 ± 0.5	world av.	18.1
$\tau \to \mu\,\nu\,\nu$	17.6 ± 0.5	world av.	17.6
$\tau \to \pi\,\nu$	10.8 ± 0.6	world av.	10.8
$\tau \to \varrho\,\nu$	22.3 ± 1.1	world av.	22.3
$\tau \to \pi\pi^0\nu$ (non res.)	$0.3 \pm 0.1 \pm 0.3$	CELLO	very small
$\tau \to \pi\pi^0\pi^0\nu$	$6.0 \pm 3.0 \pm 1.8$	CELLO	
$\tau \to \pi\pi^0\pi^0\pi^0\nu$	$3.0 \pm 2.2 \pm 1.5$	CELLO	1.0
$\tau \to \pi\pi\pi\nu$	6.4 ± 0.4	world av.	
	$9.7 \pm 2.0 \pm 1.3$	CELLO	
$\tau \to \pi\pi\pi^0\nu$	4.9 ± 0.6	world av.	
	$6.2 \pm 2.3 \pm 1.7$	CELLO	5.0
$\tau \to \pi\pi\pi\pi\pi\nu$	0.067 ± 0.030	HRS	
$\tau \to \pi\pi\pi\pi\pi\pi^0\nu$	0.067 ± 0.030	HRS	
$\tau \to K\,\nu$	0.59 ± 0.18	DELCO	0.5
$\tau \to K\,\nu + n\,\pi^0 (n > 0)$	1.13 ± 0.21	world av.	$0.8\,(K^* \text{only})$
$\tau \to KK\pi\,\nu$	$0.22^{+0.17}_{-0.11}$	DELCO	
$\tau \to K\pi\pi(\pi^0)\,\nu$	$0.22^{+0.16}_{-0.13}$	DELCO	

world average. Furthermore most experiments, with the exception of CELLO and TPC, now measure large branching ratios into 1 charged plus neutral particles. Although the discrepancy is only about 2 % (see Table 3.6) it is statistically significant due to the small experimental errors.

Let us turn to the measurements of exclusive decay channels. There is now very good agreement among the various e^+e^- experiments concerning the branching ratios for the purely leptonic decays, i.e. $\tau \to e\bar{\nu}\nu_\tau$ and $\tau \to \mu\bar{\nu}\nu_\tau$. The present world averages, neglecting the early SPEAR/DORIS results, for these branching ratios are listed in Table 3.7. The values from the individual experiments, the original references, and a more detailed discussion on the decay properties of the τ lepton can be found elsewhere (Kiesling, 1988). Note that

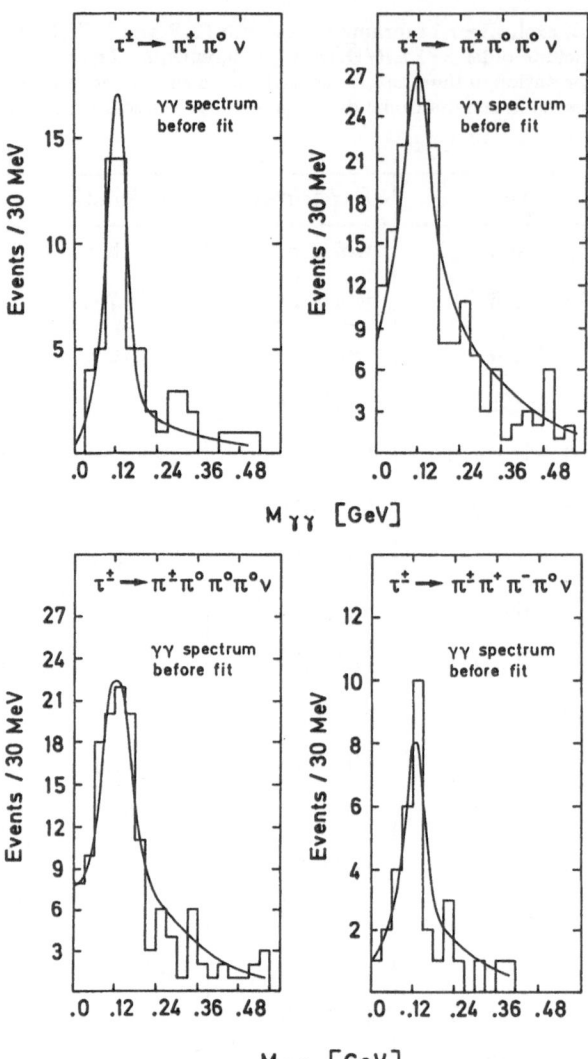

Fig. 3.27. $\gamma\gamma$ invariant masses for τ decays with 1,2, or 3 π^0's in the final state (CELLO). The curves are predictions from a detailed Monte-Carlo simulation

the ratio of muonic to electronic branching ratio, which is precisely calculable in the standard interpretation of the τ lepton, is nicely reproduced by experiment.

According to the introductory discussion the semihadronic decay channels constitute a major fraction of the τ decays. There is, however, more experimental uncertainty in the semihadronic decay modes due to the more complex final states and, in particular, due to the presence of neutral π mesons. It is evident that superior calorimetry is required to disentangle final states with different numbers of π^0's. At PETRA, a rather complete analysis on hadronic decay modes of the τ was performed by CELLO, who made extensive use of their fine grain liquid argon calorimeter. The exellent spatial resolution of the calorimeter both laterally and in depth enables separation of nearby electromagnetic and hadronic showers down to small opening angles ($\sim 1.5°$), and thus allows to restore the proper photon

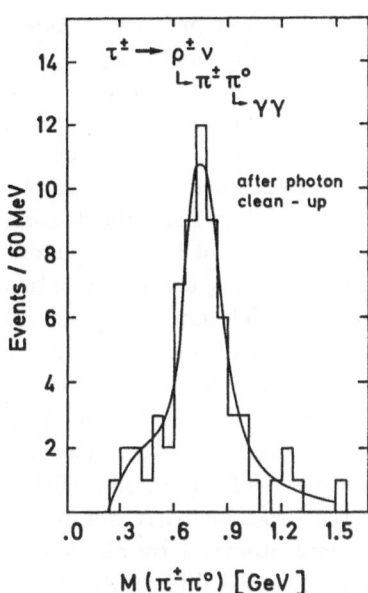

Fig. 3.28. Total invariant mass distribution from τ decays containing one charged particle and two photons (CELLO). The curves are predictions from a detailed Monte-Carlo simulation

and π^0 multiplicity. In particular, CELLO succeeded in separating semihadronic decay channels containing different numbers of neutral pions (Behrend et al., 1984). The invariant masses of the photon pairs for final states containing one, two or three π^0's are shown in Fig. 3.27. Clear π^0 signals in the various multipion final states are observed. In their analysis CELLO assume a maximum of four pions in the hadronic final state as is justified from the small branching ratio for $\tau \to 5\pi\nu$ (see Table 3.6). Furthermore, in order to take account of energetic π^0's the decay photons of which merge into a single shower, a photon multiplicity of $2n$ and $2n - 1$ is admitted for a final state containing n π^0's.

The most prominent semihadronic decay is $\tau \to \varrho\nu$. Figure 3.28 shows the invariant mass between the 2 photons from a π^0 decay and the charged particle, assumed to be a $\pi^\pm$. The curves shown in Figs. 3.27, 28 are predictions for the different final states, resulting from a detailed Monte-Carlo simulation of τ production and decay, of the detector performance, and the selection procedures for the respective τ decays. It is worth mentioning that the τ branching ratios measured by CELLO have been obtained without any assumption on the total cross section for $e^+e^- \to \tau^+\tau^-$.

The present world averages on the semihadronic τ branching ratios presently available are given in Table 3.7 as well as those measurements from CELLO which deviate in a systematic way from these averages. Original references of the measurements are contained elswhere (Kiesling, 1988). Also shown are the "safe" predictions for the semihadronic branching ratios into a stable hadron plus neutrino, and into an even number of pions based on the CVC hypothesis. Generally, quite good agreement between experiment and the predictions is found so that the standard interpretation of the τ as a sequential heavy lepton is nicely supported.

There is, however, still a problem remaining in the τ decay pattern: A closer inspection of Table 3.7 reveals a substantial fraction of the one-prong decays unexplained (about 7 % using the world averages, about 4 % using the original CELLO measurements). This discrepancy may be explained in a number of ways. One may argue that the branching ratios for the decays into one charged particle plus neutrals, including the purely leptonic decay channels, are larger, exploiting the full range allowed by the experimental errors. Thus both the theory, which uses the leptonic branching ratio as a normalisation, and the experiment would evade the problem. In doing so, however, *all* one-prong branching ratios measured so far would have to be corrected upwards, which seems to have no experimental justification. One may also argue that important one-prong decay channels have not been observed so far. This possibility has been put forward, e.g., by Gilman and Rhie (1985) who concluded that conventional decay channels such as $\tau \rightarrow \eta\pi\pi\nu_\tau$ may be substantial or that one may look for other, more unconventional. No such sources could be identified so far (see, e.g., Hitlin, 1987). Another possiblitity would be to question the large inclusive one-prong branching ratio itself, since this value is not uniquely observed by all experiments. A more detailed discussion on these problems and possible solutions can be found elsewhere (Kiesling, 1988). In any case, the decay channels of the τ lepton will remain a lively subject of research with good prospects for further clarification in the future by evaluating more data from the e^+e^- experiments. For the time being we will interpret the available experimental material as sufficient proof for a "standard" decay pattern of the τ lepton. Another important source of information comes from the lifetime of the τ lepton.

τ Lifetime and Mass of ν_τ

The measurement of the τ lifetime provides a direct determination of the coupling strength of the τ to the weak charged current. In the standard theory where the τ is viewed as a sequential lepton coupling with universal strength G to the weak charged current the width $\Gamma(\tau \rightarrow e\bar{\nu}\nu)$ is given by (see e.g. Thacker and Sakurai, 1971)

$$\Gamma(\tau \rightarrow e\bar{\nu}\nu) = \frac{G^2 m_\tau^5}{192\pi^2} = \Gamma(\mu \rightarrow e\bar{\nu}\nu)\left(\frac{m_\tau}{m_\mu}\right)^5 .$$

From the known μ lifetime T_μ and the branching ratio $B(\tau \rightarrow e\bar{\nu}\nu)$ from Table 3.7 one can calculate the expected τ lifetime T_τ

$$T_\tau = \frac{1}{\Gamma} = \frac{1}{\Gamma(\tau \rightarrow e\bar{\nu}\nu)} \cdot \frac{\Gamma(\tau \rightarrow e\bar{\nu}\nu)}{\Gamma}$$

$$= T_\mu \left(\frac{m_\mu}{m_\tau}\right)^5 B(\tau \rightarrow e\bar{\nu}\nu) \tag{3.50}$$

$$= (2.90 \pm 0.08) \times 10^{-13}[\text{s}] .$$

The error of the theoretical expectation is given by the uncertainty in the electronic branching ratio. At PEP/PETRA energies the mean flight distance of the τ before disintegration is of order 1 mm. This is large enough so that the point of decay can be distinguished on average from the point of interaction. In order to measure the τ decay length both the interaction point and the point of decay must be known. Since the bunch lengths of the colliding beams along the beam (z) direction is large ($\sigma_z \sim 1.5$ cm) compared to the average decay length, only the transverse projection is considered ($\sigma_{x,y} \sim 0.1-0.5$ mm). The beam position in x and y for the different accelerator fills is monitored by means of Bhabha events. The individual points of decay are reconstructed from three-prong decays by fitting for a common vertex of the three tracks. Early measurements of the τ lifetime are available from MARK II (Feldman et al., 1982), MAC (Ford et al., 1982a) and CELLO (Behrend et al., 1983b). More recently, TASSO (Althoff et al., 1984b), MARK II (Jaros et al., 1983, 1984), JADE (Steffen, 1984), DELCO (Klem et al., 1984), MAC (Fernandez et al., 1985; Ritson, 1986), HRS (Baranko, 1985) and ARGUS (Albrecht et al., 1987) have published accurate τ lifetime measurements, partly based on high precision vertex recognition. Recent reviews of these measurements were given by V. Lüth (1985) and D. Hitlin (1987). The new world average for the τ lifetime is (see the recent reviews by Hitlin, 1987 and Kiesling, 1988)

$$T_\tau = (3.02 \pm 0.08) \times 10^{-13} [\text{s}] \ ,$$

in very good agreement with the expectation from the standard theory based on the electronic branching ratio (3.50).

As it stands, the conclusion to be drawn from the measurements on branching ratios and lifetime of the τ is, that the data are in very good agreement with the sequential lepton hypothesis within the standard theory. Other experimental information corroborating this hypothesis, such as small upper limits on radiative decays, lepton-number-violating decays, momentum spectra of decay particles and upper limits on the τ neutrino mass, has been reviewed extensively in the literature (Feldman, 1978; Kirkby, 1979; Flügge, 1979; Tsai, 1980; Perl, 1980). Since then a new measurement of the Michel parameter for the τ decaying into $e\,\bar{\nu}\,\nu_\tau$ has been presented by CLEO (Galik, 1985) in full accord with the $V - A$ expectation of 3/4. Also the upper limit for the mass of the τ neutrino has considerably improved by virtue of the higher statistics collected by now. The upper limit for the neutrino mass is very sensitive to the Q value available in the τ decay. The rare channels with many pions or kaons are favourable for stringent mass limits due to the low Q value involved. Using their 10 events in the five prong decay channel with and without an additional π^0 HRS (Abachi et al., 1985) have obtained an upper limit of 84 MeV/c^2 at the 95% C.L. An even better limit of $m(\nu_\tau) < 35 \text{MeV}/c^2$ at 95% C.L. has been reported the ARGUS collaboration (Albrecht et al., 1987). A summary of limits on the τ neutrino mass can be found in a recent review on τ decays (Kiesling, 1988).

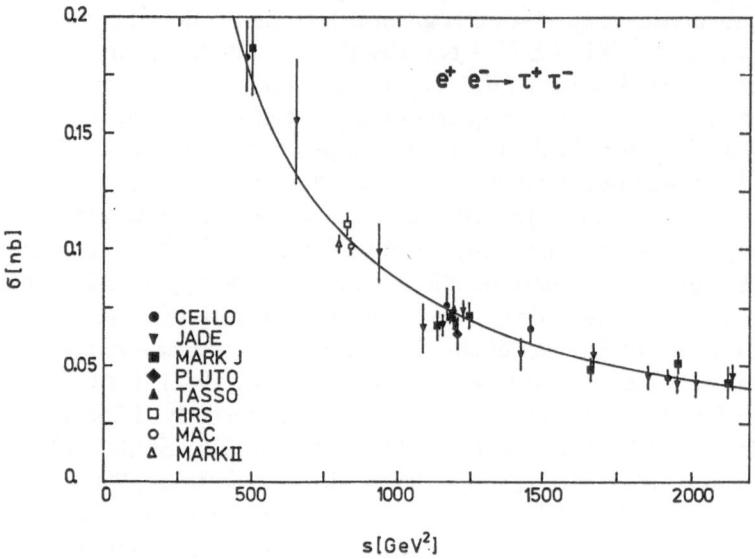

Fig. 3.29. Total cross section measurements for $e^+e^- \rightarrow \tau^+\tau^-$ as a function of the square of the centre of mass energy

Cross Section

Similar to μ pair production, the total cross section for the reaction $e^+e^- \rightarrow \tau^+\tau^-$ can be analysed in two ways. If universality of the leptonic weak couplings holds as postulated in the standard theory (see also (3.46)), no measurable deviations from the lowest order QED prediction are expected according to the discussion in Sect. 3.3. Nevertheless, measurements of the total cross section can be used to ensure that no large deviations from universality occur. If one accepts the standard theory, on the other hand, the value of the total cross section can be used to probe the electromagnetic structure of the τ.

The available measurements on $\sigma_{\tau\tau}$ from PEP and PETRA are shown in Fig. 3.29. Published data exist from HRS (Gan et al., 1985), MAC (Fernandez et al., 1985a), MARK II (Levi et al., 1983) at PEP and from CELLO (Behrend et al., 1982a), JADE (Bartel et al., 1985b), MARK J (Barber et al., 1980; Adeva et al., 1984), PLUTO (Berger et al., 1981, 1985a), TASSO (Brandelik et al., 1982; Althoff et al., 1985a) at PETRA. As for all e^+e^- processes, radiative (QED) effects are large and have to be corrected for, taking into account the experiment-specific selection algorithms for τ events. Again the Monte-Carlo method is best suited for a quantitative evaluation of these effects. Typical corrections for the total cross section are of order 35 %. The program of Berends and Kleiss (1981) is generally used, which includes radiative corrections to order α^3 for the photon exchange born term but does not give the "full" QED correction as for the reaction $e^+e^- \rightarrow \mu^+\mu^-$ (Berends et al., 1982). Recently a program has become available for τ production (Jadach, 1985) which takes into account the spin correlation of the produced τ's in presence of radiative corrections. In the nomenclature of Sect. 3.2, the program provides the reduced QED correction of order α^3. The CELLO collaboration has used this program to correct their $\tau^+\tau^-$ data for radiative effects.

All cross section measurements agree well with QED and, of course, also with the standard theory for $\sin^2\theta_W = 0.23$. Parametrising the available data on cross sections in terms of a form factor as in (3.47) leads to typical values for $\Lambda_\pm$ of order 200 GeV at the 95% C.L. (Kiesling, 1985). This means that the τ does not show any structure down to distances of $\sim 10^{-16}$ cm. This fact, though natural within the standard theory, is still remarkable in view of the τ having almost twice the mass of the proton, but a charge radius at least 3 orders of magnitude smaller.

Charge Asymmetry

Accepting the τ as a pointlike spin 1/2 lepton, the angular distribution of the reaction $e^+e^- \rightarrow \tau^+\tau^-$ should show the same charge asymmetry as the reaction $e^+e^- \rightarrow \mu^+\mu^-$. For τ pairs, however, backgrounds such as two photon processes, multihadronic events and (radiative) Bhabha scattering are large, and their suppression depends on the detector's ability to correctly identify and discriminate these final states. The contamination from Bhabha events is particularly dangerous due to their strong positive charge asymmetry. Even a small amount of Bhabha events in the τ sample may significantly reduce the observed charge asymmetry, which is expected to be negative in the standard theory.

The experimental cuts applied usually purify the τ samples to better than $\sim 95\%$. Residual backgrounds as well as radiative effects which yield a small positive charge asymmetry of approximately 1.5%, as for the μ pair production, are commonly subtracted from the angular distributions bin by bin using Monte Carlo calculations. Since the τ direction is not known various estimators are used, such as the event sphericity axis (defined as the direction in space, with respect to which the sum of the transverse momenta squared of the observed particles becomes minimal), the vector sum of the momenta of the τ decay particles , and others. These estimators approximate the τ direction to better than 10° at PEP/PETRA energies. The small smearing of the angular distribution introduced by using the decay products to define the τ direction of flight can be corrected for by Monte-Carlo methods. The systematic errors from the various background subtraction and correction procedures are generally estimated to about 1%.

Figure 3.30 shows the angular distributions for $e^+e^- \rightarrow \tau^+\tau^-$ from the PETRA experiments at $\sqrt{s}$ around 34.5 GeV. The data are from CELLO (Behrend et al., 1982b), JADE (Bartel et al., 1985b), MARK J (Adeva et al., 1984, 1986), and TASSO (Althoff et al., 1985a). Not shown are the data from PLUTO (Berger et al., 1985a) which have been taken in a somewhat more restricted angular range. Figure 3.31 shows results from the high energy running obtained by CELLO and JADE (Bartel et al., 1985b). The 1986 data at $\sqrt{s} = 35$ GeV have been evaluated by CELLO, by the JADE collaboration (Bartel et al., 1987), and by TASSO (Braunschweig, 1988). Three experiments from PEP have published differential

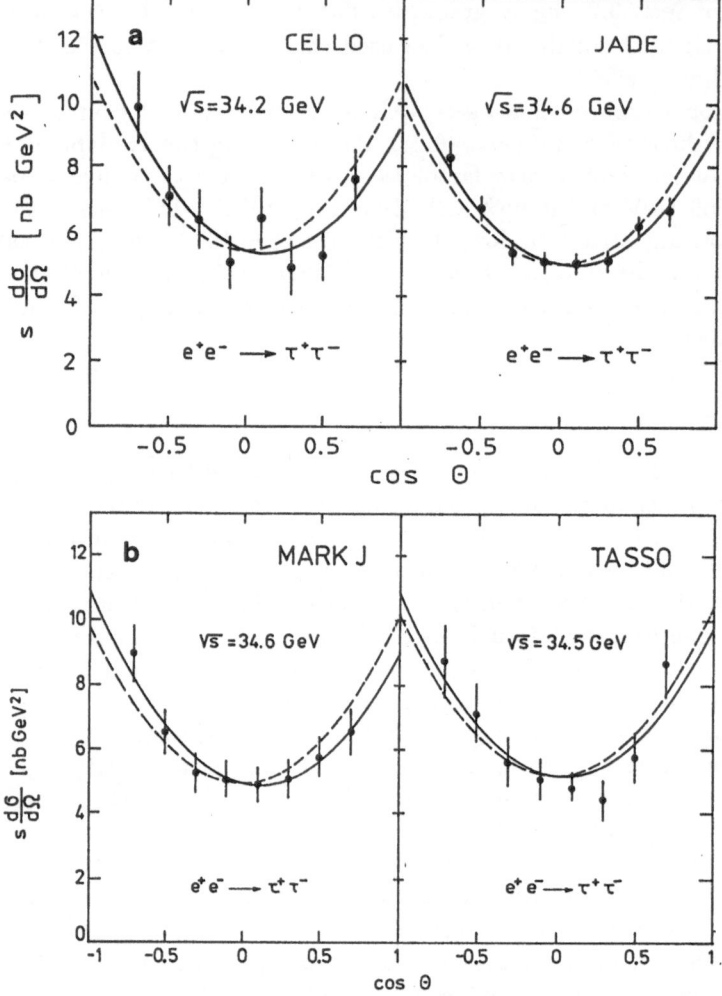

Fig. 3.30. Differential cross sections for $e^+e^- \rightarrow \tau^+\tau^-$ for the PETRA experiments (a: CELLO and JADE, b: MARK J and TASSO) around $\sqrt{s} = 34.5$ GeV. The data do not include the 35 GeV runs in 1986. The dashed curves correspond to the lowest order QED expectation. The solid curves result from fits to the data

cross sections for $e^+e^- \rightarrow \tau^+\tau^-$ so far. The MARK II collaboration (Levi et al., 1983) and HRS (Gan et al., 1985) have analysed about 100 pb^{-1} while the new data from MAC (Fernandez et al., 1985) is based on 210 pb^{-1}. The differential cross section from MAC is derived from a sample of more than 10 000 events and is shown in Fig. 3.32 together with the HRS data.

The angular distributions shown in Figs. 3.30–32 are corrected for radiative effects to order α^3. The full lines in these figures result from fits to the data while the dashed curve represents the symmetric lowest order QED expectation. Almost all experiments measure significant deviations of the charge asymmetry

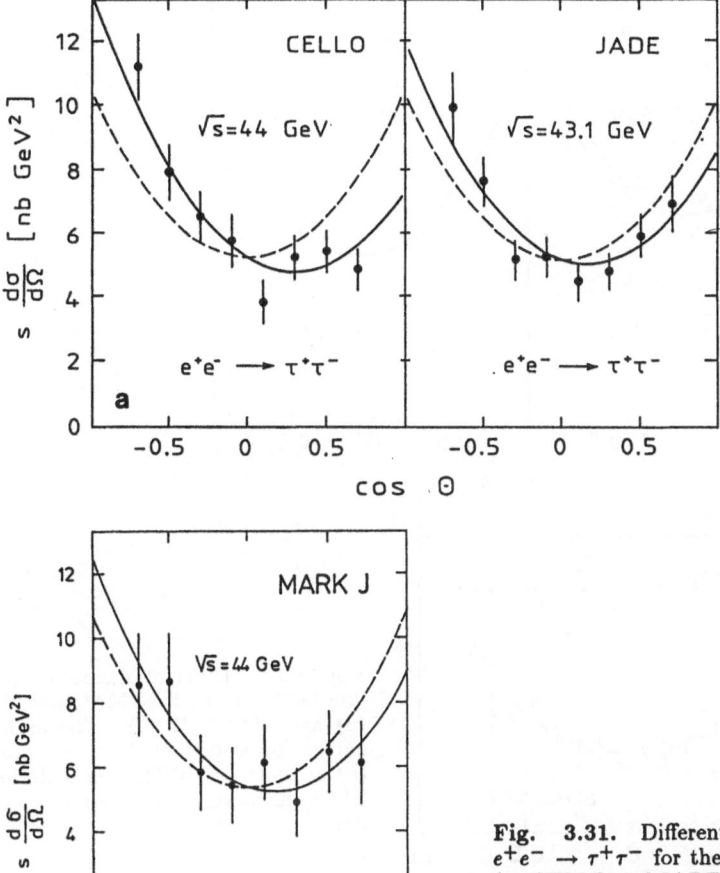

Fig. 3.31. Differential cross sections for $e^+e^- \rightarrow \tau^+\tau^-$ for the PETRA experiments (a: CELLO and JADE, b: MARK J) around $\sqrt{s} = 44.0$ GeV. The dashed curves correspond to the lowest order QED expectation. The solid curves result from fits to the data

from the QED expectation. The charge asymmetries derived from the data, corrected for reduced QED only, are listed in Table 3.8. The errors for the τ asymmetries are somewhat larger compared to the μ asymmetries (see Table 3.4), which is due to the lower statistics (generally only a fraction of the τ decays are observed) and the increased systematic uncertainties in the extraction of the τ pair event samples.

At high energies a 4.5 standard deviation effect is established by CELLO and JADE, in good agreement with the expectation from the standard theory. As a curiosity, JADE measures a positive value for the charge asymmetry at $\sqrt{s} = 38$ GeV, consistent with QED, but almost 3 standard deviations away from the expected (negative) value. CELLO, on the other hand, obtains a charge asymmetry in full accord with the standard theory at the same energy, so that the JADE

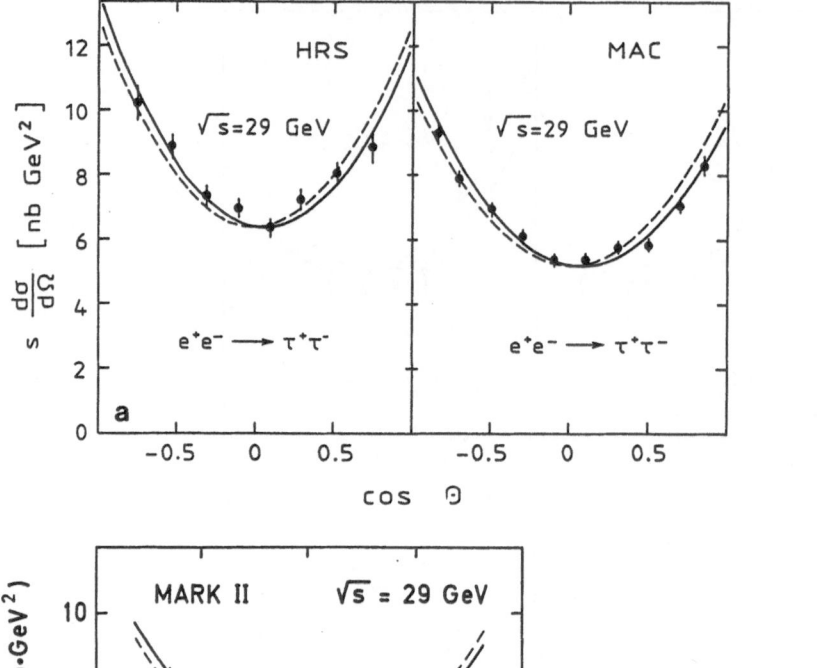

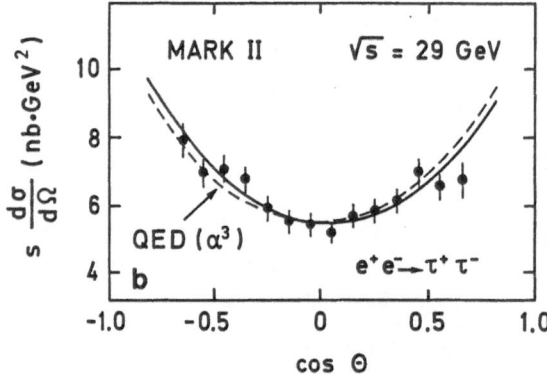

Fig. 3.32. Differential cross sections for $e^+e^- \to \tau^+\tau^-$ from MAC and HRS (a) and MARK II (b). The dashed curves correspond to the lowest order (a) or order α^3 (b) QED expectation. The solid curves result from fits to the data

measurement most likely is a statistical fluctuation. The PEP experiments now all measure deviations from QED at the better than 2σ level. MAC has obtained the most precise PEP result, with a significance of 4.2 standard deviations, in excellent agreement with the standard theory. Also shown in Table 3.8 are the averages for the PEP and PETRA energy ranges. Individual experiments have been weighted by their combined statistical and systematic errors. For similar reasons as in the case of the muon asymmetries, the systematic errors for the τ asymmetries cannot be regarded as uncorrelated between the different experiments. For the averages a mean systematic error has therefore been estimated by weighting the individual systematic errors according to the event statistics. With the combined statistical and systematic errors clear deviations from QED with a significance of 5 (PEP) and 6 (PETRA) standard deviations are observed.

First results on τ pair production from TRISTAN have been reported recently (Kamae, 1988). The measurements of the charge asymmetries from AMY, TOPAZ and VENUS are given in Table 3.9. The experimental uncertainties are largely dominated by statistics. Taking the average over the TRISTAN charge

Table 3.8. Measurements of the charge asymmetry for the reaction $e^+e^- \to \tau^+\tau^-$ from PEP and PETRA. The predictions from the standard theory are given in the last column, with $\sin^2\theta_W = 0.23$ and $M_Z = 92$ GeV, including effects due to the finite mass of the τ lepton (see also Sect. 3.6)

Exp.	$\sqrt{s}$ [GeV]	$\int \mathcal{L}\, dt$ [pb^{-1}]	$N_{\tau\tau}$	$\varepsilon_{\tau\tau}{}^*$	$A_{\tau\tau}$ [%]	A_{GSW} [%]
HRS	29	256	7372	0.28	$-4.4 \pm 1.5 \pm 0.5$	-5.7
MAC	29	210	10153	0.47	$-5.5 \pm 1.2 \pm 0.5$	-5.7
MARK II	29	100	3714	0.36	-4.2 ± 2.0	-5.7
PEP	29	566	21239	0.36	$-4.9 \pm 0.8 \pm 0.5$	-5.7
CELLO	34.2	11	434	0.52	-10.3 ± 5.2	-8.4
JADE	34.6	62	1998	0.44	$-6.0 \pm 2.5 \pm 1.0$	-8.6
MARK J	34.7	148	1401	0.13	$-10.6 \pm 3.1 \pm 1.5$	-8.6
PLUTO	34.6	42	419	0.14	$-5.9 \pm 6.8^{+0.0}_{-2.5}$	-8.6
TASSO	34.4	69	856	0.17	$-4.9 \pm 5.3^{+1.3}_{-1.2}$	-8.5
PETRA	34.6	332	5108	0.21	$-7.6 \pm 1.3 \pm 1.2$	-8.6
CELLO	35.0	87	3032	0.52	$-7.0 \pm 1.9 \pm 0.9$	-8.8
JADE	35.0	90	2900	0.45	$-8.5 \pm 2.0 \pm 1.0$	-8.8
TASSO	35.0	108.5	476	0.17	$-9.2 \pm 5.2^{+0.7}_{-3.1}$	-8.5
PETRA	35.0	285.5	6419	0.32	$-7.9 \pm 1.3 \pm 1.2$	-8.6
CELLO	38.1	9	260	0.48	$-11.8 \pm 6.2 \pm 2.7$	-10.8
JADE	38.0	12	336	0.46	$+7.5 \pm 6.3 \pm 1.0$	-10.7
PETRA	38.0	21	596	0.47	$-1.6 \pm 4.4 \pm 1.8$	-10.7
CELLO	43.8	40	824	0.45	$-16.3 \pm 3.5 \pm 1.3$	-15.2
JADE	43.7	43	913	0.47	$-17.0 \pm 3.6 \pm 1.0$	-15.1
MARK J	43.8	37.8	287	0.17	$-8.5 \pm 6.6 \pm 1.5$	-15.3
PETRA	43.8	120.8	2024	0.37	$-16.2 \pm 2.4 \pm 1.2$	-15.2

* overall efficiency for τ pairs assuming $R_{\tau\tau} = 1$.

Table 3.9. Measurements of the charge asymmetry for the reaction $e^+e^- \to \tau^+\tau^-$ from TRISTAN. The predictions from the standard theory are given in the last column, with $\sin^2\theta_W = 0.23$ and $M_Z = 92$ GeV, including effects due to the finite mass of the τ lepton (see also Sect. 3.6)

Exp.	$\sqrt{s}$ [GeV]	$A_{\tau\tau}$ [%]	A_{GSW} [%]
AMY	54.5	-36.2 ± 10.6	-27.4
TOPAZ	54.5	-20.3 ± 11.8	-27.4
VENUS	54.3	-20.0 ± 11.0	-27.1
TRISTAN	54.5	-26.0 ± 6.4	-27.4

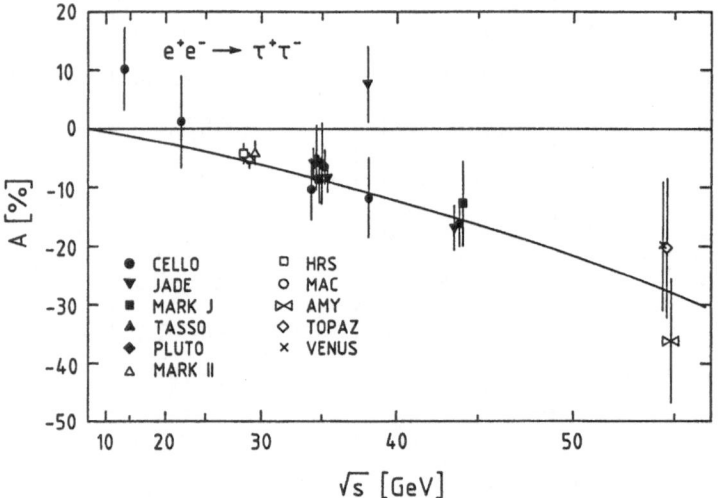

Fig. 3.33. Charge asymmetry measurements for the reaction $e^+e^- \to \tau^+\tau^-$ as function of $s = E_{\text{CM}}^2$. The curve corresponds to the lowest order expectation from the standard theory with $M_Z = 92$ GeV and $\sin^2\theta_W = 0.23$

asymmetry measurements leads to an almost 4 standard deviation effect, in excellent agreement with the theoretical expectations.

Figure 3.33 shows the available charge asymmetry measurements as a function of the square of the center-of-mass energy. The line at $A_{\tau\tau} = 0$ corresponds to the lowest order QED expectation. Also shown in the figure is the prediction from the standard theory for $\sin^2\theta_W = 0.23$ and $M_Z = 92$ GeV according to the parametrisation (3.9). Although the charge asymmetries derived by the various experiments for the τ final states are in overall good agreement with the expectation from the standard theory it is interesting to observe that for the PETRA experiments, except CELLO, the asymmetry for muons is larger in magnitude than for taus (compare Tables 3.4,8). The effect is certainly small, about one standard deviation at each of the averaged energy points, but still systematic. While this could potentially point towards a violation of universality of the weak neutral current, the accuracy of the existing data is insufficient to safely draw such a conclusion. More likely, a systematic shift of the τ asymmetry towards smaller values may be caused by residual contaminations from background with a strong positive charge asymmetry, such as radiative Bhabha scattering, deep-inelastic Compton scattering and other higher order QED processes. It is unlikely that the radiative correction programs used in the μ and τ cases which have been developed to a different level of sophistication (see Sect 3.2) could contribute to systematic differences. At any rate, as experiments improve their methods to purify their τ pair samples, the detailed comparison between μ pair and τ pair production will remain an interesting issue for testing the standard theory. Given the present τ charge asymmetry measurements shown in Tables 3.8,9, good agreeement with the predictions of the standard theory is observed.

Polarisation

Since in the standard theory the coupling of the Z^0 is different for left- and right-handed fermions, a longitudinal polarisation of the final state fermions is predicted which is proportional to the product of vector and axial vector coupling constants of the fermions involved. In contrast to the observables discussed so far, a non-vanishing longitudinal polarisation is a truely parity-violating effect. Neglecting the small purely weak terms, the polarisation $P(\theta)$ for τ leptons as a function of the polar scattering angle θ with respect to the beam axis is given by (see Sect. 3.1.3):

$$P(\theta) = -\mathrm{Re}\{g(s)\} \left(v_e a_\tau + v_\tau a_e \frac{2\cos\theta}{1 + \cos^2\theta} \right) . \tag{3.51}$$

The weak decays of the τ lepton can be used to analyse the polarisation of the produced τ's. The pertinent physics is most easily discussed with a two-body decay such as $\tau \to \pi\nu, \varrho\nu$. Assuming a standard $V - A$ charged current decay for the $\tau^\pm$ the distribution of the emission angle θ^* of the $\pi^\pm$ or $\varrho^\pm$ with respect to the $\tau^\pm$ spin direction in the τ rest system is described by (Tsai, 1971):

$$dW^\pm = \frac{1}{2} (1 \mp \alpha \cos\theta^*) \, d\cos\theta^* , \tag{3.52}$$

where α is a measure of the analysing power of the decay, related to the spin properties of the decay meson. For spin 0 mesons, such as the π meson, $\alpha = 1$, while for spin 1 mesons (e.g. ϱ and A_1) of mass m_X, α is given by

$$\alpha = \frac{m_\tau^2 - 2m_X^2}{m_\tau^2 + 2m_X^2} . \tag{3.53}$$

For the ϱ meson one therefore has $\alpha \sim 0.46$.

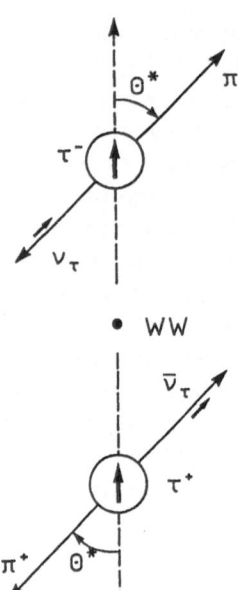

Fig. 3.34. Schematic representation of the angular correlation of the decay products of the τ in the reaction $e^+e^- \to \tau^+\tau^-$, originating from the interaction point labelled WW. The arrows attached to the particle directions denote the respective helicities

The definition of the decay angle θ^* with respect to the $\tau^\pm$ helicity axes is shown in Fig. 3.34 for the case of both τ leptons decaying into a pion and a neutrino. Since the helicities for τ^+ and τ^- are opposite in the final state (see Sect. 3.1.1) the emission angles of the decay pions with respect to the $\tau^\pm$ flight directions are correlated: Decay pions from negative (positive) τ's tend to be emitted preferentially with positive (negative) helicity along the respective τ directions of flight in the laboratory. Correspondingly, they tend to be emitted against the τ directions of flight when the τ helicities are reversed. This angular correlation therefore leads to a positive correlation of the pion laboratory momenta.

For unpolarised τ leptons the decay angular distribution is isotropic in the τ rest system resulting in a flat momentum distribution for the decay pion in the laboratory. When the τ^- polarisation is positive, i.e. the τ spin is pointing preferentially into the direction of motion of the τ^-, an excess of pions with larger-than-average momenta is expected. Correspondingly, a negative polarisation of the τ^- leads to an excess of smaller-than-average momenta. Expressing the laboratory momenta of the decay particles in units of the beam energy $\frac{1}{2}\sqrt{s}$, i.e. $x = 2p_{\mathrm{lab}}/\sqrt{s}$, and neglecting mass effects, one obtains the distribution of x by transforming (3.52), appropriately weighted by the two helicity states of the τ^- to account for a possible polarisation P, to the laboratory system (see e.g. Augustin, 1979; Goggi, 1979):

$$\frac{dW}{dx} = 1 + 2\alpha P\left(x - \frac{1}{2}\right). \tag{3.54}$$

The polarisation can thus be determined by measuring the slope of the laboratory momentum spectra for π's and ϱ's.

Muon and electron decays may also be used. However, due to the three-body nature of the purely leptonic decay the laboratory momentum distributions are somewhat more complicated. One obtains (Goggi, 1979)

$$\frac{dW}{dx} = a(x) + P\,b(x) \tag{3.55}$$

with

$$a(x) = \frac{1}{3}\left(5 - 9x^2 + 4x^3\right)$$

$$b(x) = \frac{1}{3}\left(1 - 9x^2 + 8x^3\right).$$

Although the leptonic decays have only about 1/3 of the analysis power of pions, the are still quite useful for the analysis of the τ polarisation due to their substantial branching ratios.

The expression for the polarisation $P(\theta)$ (3.51) contains two contributions. One term, independent of the production angle, is proportional to the product $v_e a_\tau$, the other term, linear in $\cos\theta$, is proportional to $v_\tau a_e$. As shown in Sect. 3.1.3, information on the vector coupling constant v_τ can be obtained by forming the polarisation asymmetry $A(P)$, which isolates the term proportional to v_τ. It is therefore evident to use the polarisation asymmetry for a measurement of the vector coupling constant.

At PEP/PETRA energies, the expectation for $A(P)$ from the standard theory is about $+1\%$, which is clearly unmeasureable at present. Still one may verify that the polarisation in τ pair production is not large (the polarisation asymmetry of the d quark, e.g., is predicted around $+18\%$, see Fig. 3.5). The CELLO experiment (Behrend et al., 1983a) was the first to analyse the produced τ leptons for a possible polarisation. For this purpose the decays $\tau \to e\,\bar{\nu}\nu,\ \mu\,\bar{\nu}\nu,\ \pi\,\nu$ and $\varrho\,\nu$ were selected and the laboratory momentum spectra for the charged decay products in each hemisphere were determined. The raw spectra receive corrections from inefficiencies in both the τ selection and particle identification procedures (muons, pions and electrons have to be distinguished by means of spectrometer and calorimeter information), as well as from radiative effects which tend to soften the spectra. The corrected laboratory momentum spectra of the four reactions, integrated over both hemispheres, are shown in Fig. 3.35.

A fit combining the eight spectra parametrised as in (3.54) and (3.55) yields a polarisation asymmetry $A(P) = (+1\pm22)\%$. The error is dominated largely by statistics. The value for $A(P)$ is clearly compatible both with zero and with the standard theory. While the obtained accuracy is still very modest this measurement of the polarisation asymmetry led to the first independent determination of the vector coupling constant of the τ lepton to the weak neutral current. The value obtained for v_τ is -0.1 ± 2.8. The standard theory limits the vector couplings v_l of the charged leptons $l = e, \mu, \tau$ within $-1 \le v_l \le +3$ (see (2.36)), and for $\sin^2\theta_W = 0.23$ a value of $v_l = -0.08$ is expected.

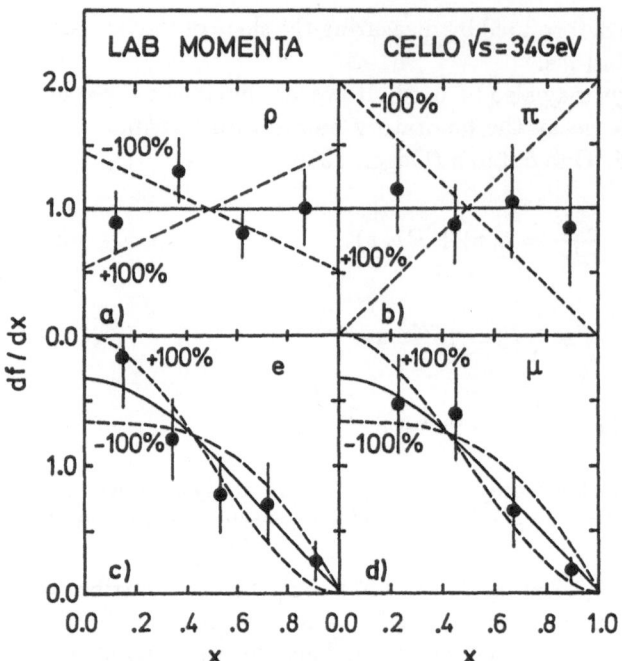

Fig. 3.35.
Distribution of laboratory momenta for the charged decay products in the reaction $e^+e^- \rightarrow \tau^+\tau^-$. The dashed curves show the prediction for complete polarisation of the produced τ leptons. The expectation from the standard theory is very close to the unpolarised case (*solid lines*)

Recently, the MAC collaboration (Maruyama, 1986) has repeated this measurement using the same decay channels as CELLO, but with considerably higher statistics. Their result at $\sqrt{s} = 29$ GeV is $A(P) = (+6 \pm 8)\%$, in agreement with the expected polarisation asymmetry of $+0.8\%$. The corresponding value for the vector coupling constant of the τ is $v_\tau = -1.04 \pm 1.24$. Also here the error is still too large to make any serious tests of universality of the leptonic vector coupling constants.

3.5 Analysis of the Lepton Data

In the preceeding chapter the available experimental information from the leptonic reactions $e^+e^- \rightarrow e^+e^-$, $\mu^+\mu^-$, $\tau^+\tau^-$ sensitive to the weak neutral current such as cross sections, charge asymmetries, and polarisations have been presented. These data will now be interpreted in the context of the standard theory. The previous discussion showed that electroweak interference has been observed beyond doubt in the charge asymmetry for μ and τ pair production, both at PEP and PETRA. The new charge asymmetry measurements from TRISTAN, although still less accurate, and the measurements from the lower PEP/PETRA energies agree quantitatively with the expectation from the GSW theory. The measured charge asymmetries were shown in Figs. 3.25,33.

A first step towards a common analysis of the leptonic channels within the framework of the standard theory may be to verify a basic assumption of the

Table 3.10. Determination of the leptonic axial vector charges from measurements of the charge asymmetries

	$a_e \, a_\mu$	$a_e \, a_\tau$
PEP	1.00 ± 0.13	0.84 ± 0.16
PETRA	1.11 ± 0.08	0.90 ± 0.11
TRISTAN	1.06 ± 0.17	0.94 ± 0.23
World data	1.07 ± 0.06	0.88 ± 0.08

theory, i.e. the equality of the weak neutral couplings v_l and a_l for the different generations of leptons ($l = e$, μ, τ). This property of the standard theory is often called "universality". The vector couplings $v_{\mu,\tau}$ can in principle be measured via the total cross section employing, e.g., (3.46):

$$R_{\mu,\tau}(s) = 1 + v_e v_{\mu,\tau} \operatorname{Re}\{g(s)\} + \frac{1}{4}(v_e^2 + a_e^2)(v_{\mu,\tau}^2 + a_{\mu,\tau}^2)|g(s)|^2.$$

As was shown in the preceeding chapter, the electroweak corrections to the total cross section, however, are very small, both in the electroweak interference part (due to the small multiplier v_e) as well as in the purely weak part (due to the small propagator term $|g(s)|^2$ well below the Z^0). Therefore it is not possible to derive meaningful measurements of the vector coupling constants $v_{\mu,\tau}$. In the case of the τ lepton, polarisation measurements yield somewhat better limits for the vector charge v_τ but no equivalent measurement is possible for the muon pairs. Thus any significant test of lepton universality using the vector charges is precluded at present.

The situation is different for the axial charges. In the expression for the charge asymmetry (see (3.48))

$$A_{\mu,\tau} = \frac{3}{4} \, a_e a_{\mu,\tau} \operatorname{Re}\{g(s)\} + \dots$$

the axial charge for μ and τ is multiplied by a_e, the purely weak terms are negligible at PEP/PETRA energies. Experimentally, a_e is known to be -1 within small errors from neutrino electron scattering (see Chap. 4), as expected in the standard theory. Table 3.10 shows the values for the product $a_e \, a_l$ ($l = \mu$, τ) derived from the average charge asymmetries listed in Tables 3.4−9. The errors given in Table 3.10 include both the statistical and systematic uncertainties of the measurements. In the calculations the Glashow scheme (3.9) has been used for the propagator term, with $M_Z = 92$ GeV and $\sin^2 \theta_W = 0.23$ as usual. Taking into account the results from neutrino electron scattering one can safely set $a_e = -1$ and determine the axial vector charges for the muon and the tau. Since the values for the mass of the Z^0 and for $\sin^2 \theta_W$ have experimental uncertainties

Table 3.11. Determination of the leptonic weak charges from fits to the total and differential cross sections of the individual leptonic final states.

Lepton l	$v_e v_l$	$a_e a_l$
e	0.079 ± 0.113	1.525 ± 0.425
μ	-0.033 ± 0.076	1.043 ± 0.056
τ	-0.031 ± 0.118	0.944 ± 0.093

(see Chaps. 4 and 6), an additional error for the axial charges from these sources has been computed. One obtains

$$a_\mu = -1.07 \pm 0.07 \pm 0.03$$

$$a_\tau = -0.90 \pm 0.10 \pm 0.03 \ . \tag{3.56}$$

From PETRA alone one obtains $a_\mu = -1.13 \pm 0.08 \pm 0.03$ and $a_\tau = -0.93 \pm 0.13 \pm 0.03$. The various determinations of the axial charges are in agreement with the standard theory, which predicts $a_l = -1$, within one standard deviation. The comparison of the axial vector coupling constants for e (from ve scattering), μ and τ (from $e^+ e^-$ annihilation) impressively demonstrates the universal character of the weak interaction as formulated in the standard theory.

Instead of analysing total cross sections and charge asymmetries separately, one can directly fit the differntial cross sections for the individual leptonic reactions to the expectations of the standard theory, as functions of the weak charges. In such fits the product of the vector coupling constants $v_e v_l$ and axial-vector coupling constants $a_e a_l$ ($l = \mu, \tau$) serve as free parameters. For the case of Bhabha scattering, the differential cross sections will yield the same information for $l = e$. The errors will clearly be much larger than those from the other leptonic channels since the expected effects are small. Comparison of the various products of coupling constants, however, will further test universality.

Since not all experiments have published differential cross sections for μ and τ pair production, measurements on R_l and on A_l will be used instead. The vector and axial-vector charges for the different lepton channels, determined in fits to all data presented earlier, are shown in Table 3.11. In these fits the overall systematic errors of the differential cross sections (usually dominated by the normalisation uncertainty) were taken into account in the following way: The normalised cross section R of a given experiment was regarded as measured with the quoted systematic error δ_R and was allowed to vary in the fit, contributing the quantity $(1 - R)^2/\delta_R^2$ to the overall χ^2. As can be seen from Table 3.11, both vector and axial vector charge products calculated independently for the three leptonic channels show good agreement with the assumption of universal coupling of the lepton families to the weak neutral current.

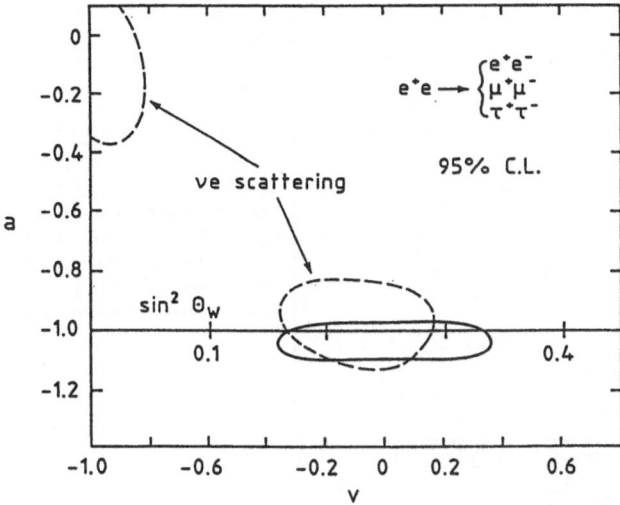

Fig. 3.36. 95% C.L. contour in the $v - a$ plane from the e^+e^- data. Only one quadrant of the plane is drawn. Note that the mirror solution centered at $v = 0, a = +1$ is off-scale. Also drawn are the corresponding contours from νe scattering. The expectation from the standard theory is indicated by the horizontal line at $a = -1$, showing the dependence on $\sin^2 \theta_W$

We next proceed to a joint analysis of all leptonic data, now assuming universality of the weak charges. The analysis will be carried out in the following steps:

— Determine the products v^2 and a^2 in a combined analysis from the total and differential cross sections for the three leptonic final states.
— Use the Weinberg-Salam scheme (3.10), use $v = -1 + 4 \sin^2 \theta_W$, $a = -1$, and fit for $\sin^2 \theta_W$ and ϱ as free parameters. Recall that this will test the prediction of $\varrho = 1$ in the standard theory.
— Use the Glashow scheme (3.9) and determine $\sin^2 \theta_W$ and M_Z.

A combined fit for the leptonic vector and axial-vector charges v^2 and a^2 using all data gives the results

$$v^2 = 0.041 \pm 0.057$$

$$a^2 = 1.004 \pm 0.056 \, ,$$

(3.57)

where the quoted error contains both statistical and systematic errors of the data, added in quadrature. The quality of the fit is good, with a χ^2 per degree of freedom of 0.601. The standard theory prediction is $a^2 = 1$ and, using $\sin^2 \theta_W = 0.23$, $v^2 = 0.006$, in excellent agreement with the data. The 95% C.L. contours in the v, a plane resulting from the fit are shown in Fig. 3.36. Since only the squares can be determined, four solutions exist for the pair (v, a) of coupling constants at $v \sim \pm 0.2$, $a \sim \pm 1$. Due to the large error of v^2 this ambiguity reduces to a twofold one for the 95 % C.L. contours. The standard theory selects the solutions with the negative axial-vector charge. From e^+e^- data alone no choice can be made, but with the help of neutrino electron scattering data (see Chap. 4) the non-standard theory solution ($a \simeq +1$) is clearly eliminated.

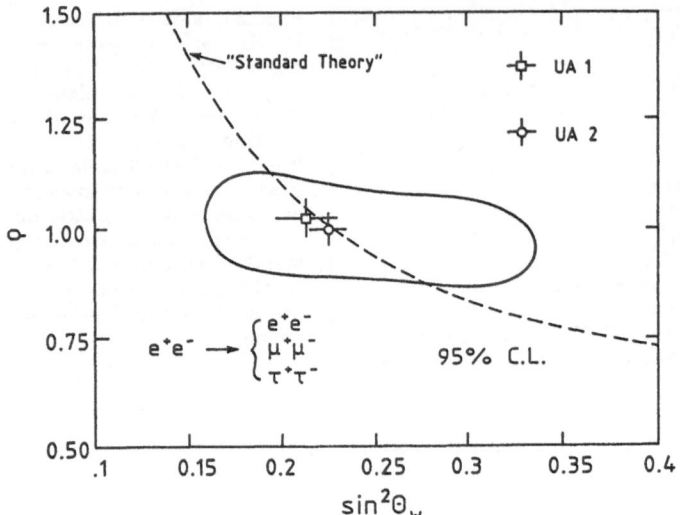

Fig. 3.37. 95% C.L. contours in the $\varrho - \sin^2 \theta_w$ plane from all e^+e^- data. The dashed line represents the relation between $\sin^2 \theta_w$ and ϱ for a fixed mass of the Z^0 ($M_Z = 92\,\text{GeV}$) given by the standard theory (see (2.41)). Also shown are the measurements and 1 σ errors of the proton-antiproton experiments

Assuming now the canonical expressions for v and a (see Table 2.1) one may choose either of the propagator terms (3.9, 10) describing Z^0 exchange to determine the free parameters of the standard theory. Working in the Weinberg-Salam scheme (3.10), the free parameters are the relative strength ϱ of the neutral current with respect to the charged current, and the Weinberg angle $\sin^2 \theta_w$, where the mass of the Z^0 has been expressed in terms of ϱ and known constants by means of (2.41). A fit to the data yields the result ($\chi^2/$ degree of freedom $= 0.600$)

$$\sin^2 \theta_W = 0.209 \pm 0.031 ,$$

$$\varrho = 1.003 \pm 0.053 .$$

When the Weinberg angle is set to $\sin^2 \theta_W = 0.230 \pm 0.005$, a one parameter fit for ϱ produces the value

$$\varrho = 0.997 \pm 0.040 \pm 0.010 ,$$

where the second error is due to the uncertainty in $\sin^2 \theta_W$ (see Chaps. 4 and 6). The overall strength of the leptonic weak neutral current thus is in nice agreement with the assertion of a minimal Higgs sector in the standard theory. The 95% C.L. contour in the (ϱ, $\sin^2 \theta_W$) plane is shown in Fig. 3.37.

Finally, the parametrisation (3.9) may be used to simultaneously determine M_Z and $\sin^2 \theta_W$ from the e^+e^- data. Here it is expected that the resultant uncertainty on $\sin^2 \theta_W$ will be minimal due to its appearance in the denominator of the propagator term. A fit to all e^+e^- data yields ($\chi^2/$degrees of freedom $= 0.926$)

$$\sin^2 \theta_W = 0.195 \pm 0.017$$

$$M_Z = 89.2 \pm 2.7 \; [\text{GeV}/c^2] .$$

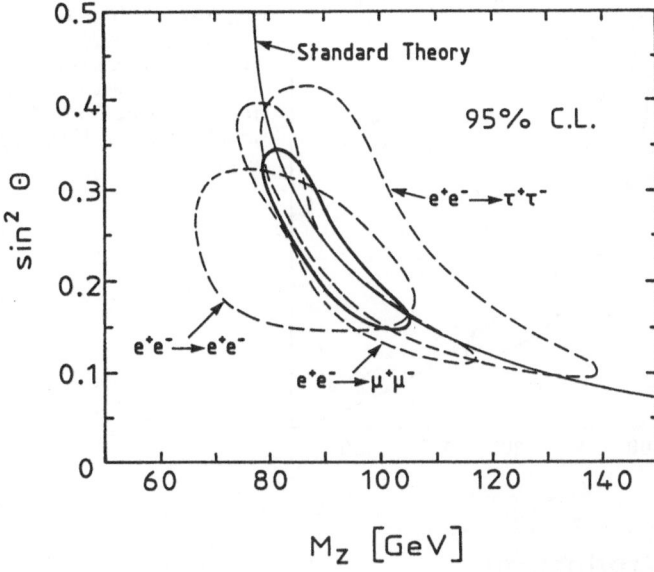

Fig. 3.38. 95% C.L. contours in the $\sin^2\theta_W - M_Z$ plane from all e^+e^- data. The contours of the individual leptonic channels are shown as dashed lines. Also given is the relation between $\sin^2\theta_W$ and M_Z as predicted in the standard theory (see (2.40))

The 95% C.L. contours in the $(\sin^2\theta_W, M_Z)$ plane are shown in Fig. 3.38. Both the values for $\sin^2\theta_W$ and M_Z agree within two standard deviations with the high precision determinations from neutrino scattering ($\sin^2\theta_W = 0.230 \pm 0.005$, see Chap. 4) and the collider experiments ($M_Z = 92$ GeV, see Chap. 6), respectively. If one assumes the value of the Weinberg angle from the neutrino experiments the mass of the Z^0 is predicted by the e^+e^- data as

$$M_Z = 90.3 \pm 2.3 \pm 1.0 \text{ [GeV]} \quad (\sin^2\theta_W = 0.23 \pm 0.005).$$

Alternatively assuming the Z^0 mass from the collider experiments the Weinberg angle from e^+e^- data is determined to be

$$\sin^2\theta_W = 0.216 \pm 0.013 \pm 0.008 \quad (M_Z = 92 \pm 1.8 \text{ [GeV]}).$$

The second error in the above determinations is systematic and due to the uncertainty of the Weinberg angle and the neutral boson mass as quoted. Figure 3.39 shows the relation between the mass of the Z^0 and $\sin^2\theta_W$, including the radiative corrections from (2.42), together with the 95 % confidence level contour from all e^+e^- data, and the measurements from the $\bar{p}p$ experiments (see Chap. 6) for comparison. The contour is the same as in Fig. 3.38. Note that the $\bar{p}p$ error bars correspond to one standard deviation only.

The e^+e^- data on leptonic final states have been shown to strongly support the underlying ideas of the standard theory. They furthermore yield measurements for the basic parameters of the model which are now becoming comparable in accuracy with other dedicated experiments. Apart from the leptonic channels the quark pair production from e^+e^- has entered a state where significant contributions to tests of the standard theory are possible. These will be discussed in the subsequent section.

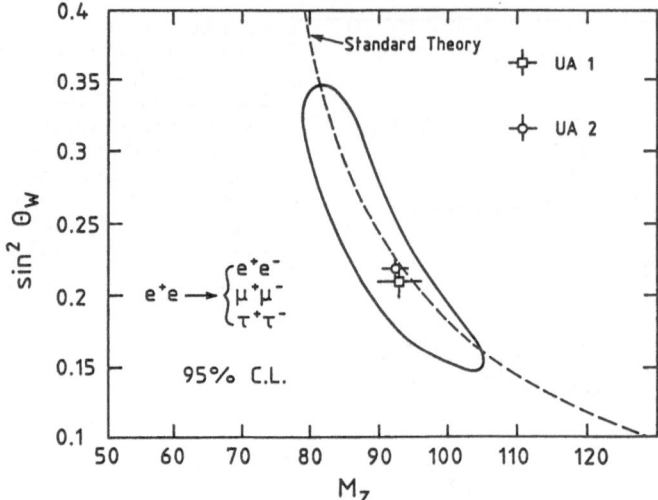

Fig. 3.39. $\sin^2 \theta_W$ vs. M_Z in the standard theory (radiative corrections are included) together with measurements from e^+e^- annihilation and the $\bar{p}p$ experiments

3.6 Quark Pair Production

Electron-positron annihilation into hadrons is well-described in many aspects by the quark-parton model where the elementary process is $e^+e^- \to \bar{q}q$. According to quantum chromodynamics (QCD), the final state quarks may radiate gluons, which lead to three (multi) jet event topologies, provided the gluon energy is sufficiently high. The corresponding three jet events diagram for the lowest order QCD process, involving one gluon, is shown in Fig. 3.40. For such processes perturbative methods (Feynman rules) exist, which enable calculations of the scattering amplitudes at the parton level. The subsequent fragmentation of the partons into hadrons is thought to proceed via "soft" QCD processes which, due to the non-Abelian nature of the colour gauge group of QCD, cannot be calculated perturbatively. Therefore various fragmentation schemes are employed to model the hadronisation process of the primary quarks and gluons, the "decay products" of which are finally measured in the detector. Most widely used are the independent fragmentation scheme (Field and Feynman, 1978) and the colour string fragmentation scheme (Andersson et al., 1980). Each scheme incorporates a set of free parameters which have to be adjusted in order to reproduce experimental distributions sensitive to the fragmentation process.

Experimentally, multihadronic events can be readily separated from the background, in particular using 4π detectors: Requiring a minimal number of charged

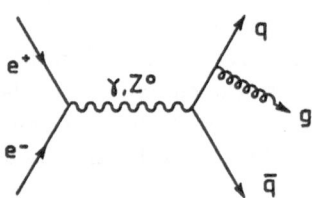

Fig. 3.40. Lowest order Feynman diagram for the reaction $e^+e^- \to q\bar{q}$ + gluon

particles (typically ≥ 6) with approximate charge, momentum, and energy balance effectively suppresses the low multiplicity leptonic reactions as well as beam-gas and cosmic ray events. Two-photon-induced multihadronic events are eliminated by requiring a minimal visible energy (typically $\geq 0.5\sqrt{s}$). The residual background from the above sources is of order $1 - 2\%$. Detector acceptances, event selection efficiencies and radiative corrections are usually determined by realistically simulating the response of the experiments to the relevant reactions. The programs used by the different experiments to estimate the radiative corrections in e^+e^- annihilation into hadrons, especially initial state radiation, are generally based on the work of Bonneau and Martin (1971), Berends and Komen (1976), Berends and Kleiss (1981, 1981a), and Berends et al. (1982).

Some of the electroweak effects in hadronic final states expected on the basis of the GSW theory, e.g. the charge asymmetry and the polarisation, are sizeable in the PEP/PETRA range only for specific $\bar{q}q$ intermediate states (see Sect. 3.1). In order to enrich or isolate events containing quarks with a specific flavour and charge one has to introduce selection criteria for the hadronic events which sometimes severely cut into the statistics. These cuts are somewhat dependent on the fragmentation scheme assumed and may therefore introduce systematic effects which have to be carefully controlled in absence of a complete knowledge of the fragmentation process. Enhancements of specific quark final states has been achieved so far only for the heavy quarks c and b. There has been considerable experimental progress recently, and the determination of the weak charges of individual quarks is slowly approaching the precision of those of the leptons.

The least model-dependent quantity for tests of the predictions of the standard theory is the total cross section, where details of the assumed fragmentation model are unimportant. Measurements of the total cross section has been used to search for the missing t quark, e.g. via the production of toponium resonances which, as is well known, has not met success so far. Most importantly, the total cross section can be analysed for weak contributions. Here, in contrast to the lepton case, the weak effect is already around 5 % at PETRA's top energy, rising steeply towards the Z^0. An additional complication, however, arises from the QCD contributions (gluon emission) to the total cross section, which are still larger than the weak effect for most of the PETRA range, and in particular for PEP. At TRISTAN, on the other hand, weak effects already dominate the part of the cross section above the naive quark-parton-model expectation.

3.6.1 Total Cross Section

In the simple quark-parton model the ratio $R_0 = \sigma_{\mathrm{had}}/\sigma_{\mu\mu}$ of the total hadronic cross section in lowest order electroweak interaction to the lowest order purely electromagnetic muon pair cross section is given by (see (3.25))

$$R_0(s) = 3 \sum_q \left[Q_q^2 - v_e Q_q v_q \, \mathrm{Re}\{g(s)\} + \frac{1}{4}(v_e^2 + a_e^2)(v_q^2 + a_q^2)|g(s)|^2 \right]. \quad (3.58)$$

The sum is over the quark flavours accessible at the center of mass energy $W =$

$\sqrt{s}$. This expression for the hadronic continuum cross section will serve as the basis for the following discussion of effects expected as the centre of mass energy for e^+e^- annihilation into hadrons approaches the Z^0 peak.

Search for the t-Quark

The standard theory demands a charge $Q = +2/3$ partner of the b quark, the t quark, to complete the left-handed weak isodoublet of the third quark generation. In e^+e^- annihilation a new heavy quark Q will show up with unique signatures: Below threshold ($\sqrt{s} < 2M_Q$), $Q\bar{Q}$ bound states will lead to narrow enhancements in the total hadronic cross section, as is well known from the charmonium and bottomonium systems. These resonances can be looked for by scanning through the available energy range in steps determined by the machine resolution ΔW, typically 30 MeV at PETRA. Since the width Γ_{tot} of such a resonance is expected to be small compared to ΔW, the cross section enhancement will depend on the actual value of the beam width. This dependence is removed by integrating the resonance cross section over a sufficiently large energy interval around the expected mass M of the $Q\bar{Q}$ bound state. The integrated hadronic cross section due to the resonant state, normalised to the lowest order muon pair cross section, is given by

$$\int R(W)\, dW = \frac{9\pi}{\alpha^2} \frac{\Gamma_{ee}\Gamma_{\text{had}}}{\Gamma_{\text{tot}}}\,, \tag{3.59}$$

where Γ_{ee} and Γ_{had} are the partial widths of the $Q\bar{Q}$ state into e^+e^- and hadrons, respectively. Note that the integral (3.59) is an *additional* contribution over the base line cross section given in (3.58), This integral, or limits thereof, can be calculated from measurements of R_{had} and can then, using assumptions on the partial width Γ_{had}, be converted into a limit for the electronic width Γ_{ee}, responsible for the excitation of a new heavy quark in e^+e^- annihilation.

The branching ratio $B_{\text{had}} = \Gamma_{\text{had}}/\Gamma_{\text{tot}}$ for heavy quarks is expected around 80%, based on measurements of semileptonic decays of b quarks (see below). For heavy quarks with charge Q, Buchmüller and Tye (1980) have estimated the partial width into electrons $\Gamma_{ee} \sim 10.1\, Q^2$ [keV], which yields for the 2 possible quark charges

$$\Gamma_{ee}(+2/3) \approx 4.5\ [\text{keV}]$$
$$\Gamma_{ee}(-1/3) \approx 1.1\ [\text{keV}]\,. \tag{3.60}$$

Searches for the t quark at PETRA have been published by CELLO (Behrend et al., 1981b, 1984d, 1987), JADE (Bartel et al., 1981, 1983, 1985), MARK J (Adeva et al., 1983a, 1986a), PLUTO (Berger et al., 1980; Criegee and Knies, 1982) and TASSO (Brandelik et al., 1982; Althoff et al., 1984a), all with negative results up to the top end of PETRA. The 95% C.L. upper limits for $\Gamma_{ee} \cdot B_{\text{had}}$ for the individual experiments are < 2.9 keV, the combined upper limit in the PETRA range up to 46.8 GeV is

$$\Gamma_{ee} \cdot B_{\text{had}} < 0.9 \text{ [keV]} . \tag{3.61}$$

While the production of the t quark is excluded in the PETRA energy range from cross section measurements alone, a fourth quark with $Q = -1/3$ is still compatible with the above limit on Γ_{ee}. When analysing the topology of multi-hadronic events, however, the PETRA experiments also exclude the production of heavy $|Q| = 1/3$ quarks with mass below 22 GeV. Since the heavy quark pair is almost at rest in the laboratory system, an excess of spherical events is expected, irrespective of the quark charge. No such excess is seen. It is interesting to note that the new total cross section measurements at TRISTAN are not yet precise enough to also exclude a new $|Q| = 1/3$ quark beyond the PETRA range (Kamae, 1988).

The non-observation of the t quark at PETRA does not in any way conflict with the standard theory, which does not predict the quark masses. Limits on the t quark mass, however, can be derived using models in the charged current sector: Recent analyses on CP violation of the $K^0 - \bar{K}^0$ system in the framework of the standard six flavour model of quarks with mixing according to the CKM matrix, expect the top quark mass far beyond the reach of PETRA (see e.g. Buras, 1988). In fact, the combination of the large experimental lifetime of the b quark of about 1.1×10^{-12} s, the small relative branching ratio of the weak b quark decays into u and c quarks (< 0.10), and the small value for the CP violation parameter ε'/ε make a top mass below 50 GeV rather unlikely. Including the new measurements on the $B^0 - \bar{B}^0$ mixing (Albrecht et al., 1987a) can be used to limit the top mass from above to less than about 120 GeV. On the other hand, these calculations depend on theoretical input describing the non-perturbative aspects of the K^0 decay amplitudes with some poorly known parameters such as the bag parameter B characterising the effective short-distance interaction of the quarks inside the K meson. One thus may still look forward to a possible discovery of the t quark either in $\bar{p}p$ interactions at the CERN collider or at the next generation of e^+e^- / ep machines now under construction.

Weak Contributions to the Continuum

In order to allow a meaningful comparison with experiment in the continuum away from the resonances, the naive quark-parton model expectation (3.58) for the cross section of e^+e^- annihilation into hadrons needs some refinements and modifications. These are the mass effects of heavy quarks and, more importantly, the QCD corrections to the naive quark-parton model. The changes are quite important, in particular since they introduce some energy dependence into the normalized total cross section, which partly hides the genuine weak contribution one is trying to measure.

At PEP/PETRA energies heavy b quarks are produced, the masses of which are not negligible in the entire energy range. Therefore additional kinematic factors related to the velocity $\beta = (1 - 4m_q^2/s)^{1/2}$ of the final state quark with mass m_q must be included. These factors are different for the purely vector, purely axial-vector, and the vector axial-vector part of the coefficients C_S and

C_A given in (3.16, 17). Taking into account the quark mass terms, the electroweak cross section R_0 (no QCD corrections yet) now reads:

$$R_0(s) = 3 \sum_q \left[\frac{\beta}{2}(3 - \beta^2)C_{VV}^q + \beta^3 C_{AA}^q \right],$$

(3.62)

with

$$C_{VV}^q = Q_q^2 - v_e Q_q v_q \operatorname{Re}\{g(s)\} + \frac{1}{4}(v_e^2 + a_e^2)v_q^2|g(s)|^2$$

$$C_{AA}^q = \frac{1}{4}(v_e^2 + a_e^2)a_q^2|g(s)|^2.$$

(3.63)

Furthermore, corrections due to the strong interaction in the final $q\bar{q}$ state have to be taken into account (see Fig. 3.40, where only one of the possible lowest order diagrams is shown). In the framework of QCD gluons can be emitted from a quark with a probability proportional to the strong coupling constant α_s. Taking into account higher order loop corrections, the effective coupling constant becomes a "running coupling constant" $\alpha_s(Q^2)$, where Q^2 now denotes the square of a momentum transfer, e.g. between the final state quarks. Starting from an arbitrary (renormalisation) scale μ the running coupling constant $\alpha_s(Q^2)$ can be obtained at any other scale Q^2 by the expression

$$\alpha_s^{-1}(Q^2) = \alpha_s^{-1}(\mu^2) + \frac{33 - 2n_f}{12\pi} \ln\left(\frac{Q^2}{\mu^2}\right),$$

(3.64)

where n_f is the number of quark flavours. An important difference may be noted comparing (3.64) with the similar expression for the effective fine structure constant α (see (3.41)): Due to the non-Abelian character of the $SU(3)$ gauge group of QCD, α_s decreases as the momentum transfer Q^2 for the quark increases (asymptotic freedom). Therefore, with rising centre of mass energies, one expects a decreasing contribution from gluon radiation. In second order QCD (see Dine and Sapierstein, 1979; Chertyrkin et al., 1979; Celmaster and Gonsalves, 1979) R is changed to

$$R(s) = R_0(s) \left[1 + \frac{\alpha_s}{\pi} + C\left(\frac{\alpha_s}{\pi}\right)^2 \right].$$

(3.65)

In the modified minimal subtraction ($\overline{\text{MS}}$) scheme as introduced by Bardeen et al. (1978) $C \approx 1.41$, resulting in a positive correction of the total cross section of $\Delta R/R \approx 0.05$. Thus knowing α_s at some (low) energy, the s dependence of the QCD correction is fixed and the electroweak contributions to the total cross section can be measured in principle. As can be seen from (3.58) and Table 2.1, the electroweak interference term is proportional to the sum of the charge-weighted vector coupling constants v_q multiplied by v_e and will lead to a reduction in the total cross section as long as $\sin^2\theta_W$ is smaller than $1/4$. However, the interference term is overwhelmed by the positive, purely weak

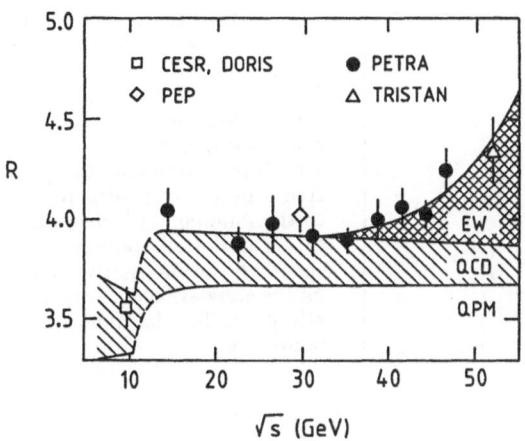

Fig. 3.41. Measurement of the total cross section for $e^+e^- \rightarrow$ hadrons in units of $\sigma_{\mu\mu}$ from all e^+e^- experiments. The expectations from the quark-parton model (QPM), including QCD corrections (QPM + QCD), and additional weak contributions according to the standard theory (EW) are indicated

term in the upper PETRA energy range. With $\sin^2\theta_W = 0.23$ one expects, at $\sqrt{s} = 45$ GeV, an increase of $\Delta R/R \approx 0.06$, in the same order of magnitude as the QCD effect.

Precise total cross section measurements over a wide energy range have been carried out by the PETRA experiments cited above and by the HRS (Bender et al., 1985) and MAC (Fernandez et al., 1985b) collaborations at PEP running at 29 GeV. First cross section measurements from TRISTAN have also been presented at the Munich Conference (Kamae, 1988). Figure 3.41 shows the available cross section data, normalised by $\sigma_{\mu\mu}$ in the usual way, averaged over the various experiments and over reasonably large energy intervals for better display. The data clearly show a rise near the top end of PETRA as expected from the purely weak terms due to Z^0 exchange in the reaction amplitude. In the precise determination of the total cross sections, the radiative corrections have become an important source of uncertainty. The data are not fully corrected for higher order radiative effects. Second and higher order corrections are necessary to test the theory at the desired level. With the presently available Monte Carlo programs, which describe initial state Bremsstrahlung, vertex corrections and vacuum polarisation, but no final state radiation or higher order photon emission, an uncertainty of order 2 % at the highest PETRA energies can be expected.

The sizes of the weak and QCD effects described are small and the systematic errors involved in the cross section determinations from individual experiments are of the same order of magnitude, certainly not much smaller. Therefore the data of all experiments should be combined to obtain a sufficient precision in R as a function of s and thus to be able to seriously test the standard theory. Within the framework of the standard theory, the free parameters are the Weinberg angle, the mass of the Z^0, and the strong coupling constant α_s. Setting the Z^0 mass to the value measured by the $p\bar{p}$ experiments, the remaining parameters can be determined in a simultaneous fit to the cross section data. In doing so, however, one has to realise the strong correlation of α_s with the absolute normalisation of the total cross section. Furthermore, the experimental input used is not entirely uncorrelated. Correlations are in fact expected, e.g. due to com-

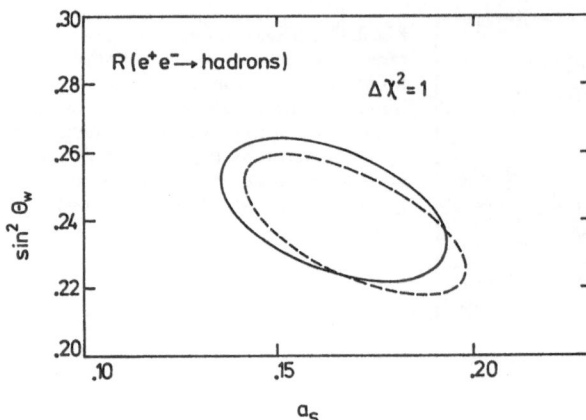

Fig. 3.42. Error contours for $\sin^2\theta_W$ vs. α_s obtained in a fit to the relative cross section R for $e^+e^- \to$ hadrons, including correlations among the various contributing experiments. The solid line corresponds to an increase in χ^2 by 1 with respect to the minimum χ^2 of the best fit. The dotted contour corresponds to an alternative model for the correlation of the systematic errors for the different experiments

mon normalisation uncertainties within a given experiment, due to the common neglect of higher order radiative corrections etc.. In a recent analysis, CELLO (Behrend et al., 1987) have taken into account these correlations, introducing a full error correlation matrix in the fitting procedure. The results from these fits for α_s and $\sin^2\theta_W$ using all PEP/PETRA cross section data above $\sqrt{s} = 10$ GeV are

$$\alpha_s(\sqrt{s} = 34 \text{ [GeV]}) = 0.16 \pm 0.03$$

$$\sin^2\theta_W = 0.236 \pm 0.020 \,.$$

In the above errors no common systematic uncertainty of the contributing experiments has been included. Setting such a hypothetical error to 1 % (recall: the uncertainties due to the radiative corrections were about 2 %) will change $\sin^2\theta_W$ by 0.005. The value for $\sin^2\theta_W$ is in nice agreement with the result from the purely leptonic reactions discussed in the previous section. It should be mentioned that the fit finds a second solution at $\alpha_s = 0.19 \pm 0.03$ and $\sin^2\theta_W = 0.56 \pm 0.02$. This solution is characterised by a large negative interference term and a large positive purely weak term, due to the large vector coupling constant v_e for this value of $\sin^2\theta_W$, which is, however, ruled out by the neutrino experiments. To indicate the correlation between $\sin^2\theta_W$ and α_s, the error contours from the fit to $\sin^2\theta_W$ and α_s with two different assumptions on the correlation of the systematic errors between the various contributing experiments are shown in Fig. 3.42. The correlation coefficient is 0.49 for solid error contour in the figure. The analysis has been recently updated (D'Agostini et al., 1988) to include QCD corrections to R up to third order, which have been now calculated (Gorishny et al., 1988), the result on $\sin^2\theta_W$, however, does not change.

At present, the $\sin^2\theta_W$ determination from the total hadronic cross section measurements is less precise than the one from the leptonic channels. Knowing α_s with better precision certainly would reduce the error on $\sin^2\theta_W$ from the cross section measurements. Observables characterizing the jet structure of the hadronic final states can indeed be used to measure α_s with a much higher (statistical) precision. As is well known, albeit not entirely undisputed, a strong dependence of α_s on the fragmentation scheme used (see, e.g., Kiesling, 1982; Behrend et al., 1983a, 1984c) has been observed when analysing shape variables of multi-jet events. In the cross section analysis, on the other hand, the fragmentation models only enter in the overall acceptance calculations for hadronic events. Uncertainties in the details of the fragmentation model used to calculate the event selection efficiencies are therefore not expected to introduce additional systematic erros in the determination of the Weinberg angle.

3.6.2 Charge Asymmetries for Light Quarks

As discussed in Sect. 3.1, the electroweak charge asymmetry expected in the standard theory strongly depends on the flavour of the quarks produced. For multihadronic events which are not flavour-separated, the expected charge asymmetry is largely reduced and even sign-reversed with respect to the lepton final states (see (3.23)). In addition, since only the fragmentation products are measured in a detector, certain algorithms have to be employed in order to define or tag the charge of the parent quark. The efficiencies and possible biases of these methods can only be determined by Monte Carlo techniques, which necessarily have to use models for the fragmentation process. One natural assumption to tag the charge of the parent quark is the concept of the "leading particle" (Field and Feynman, 1978) which attributes a larger probability for a fast (or leading) hadron to contain the primary quark as compared to a slow hadron. In terms of the rapidity η_i of the charged particles i, the jet charge Q_{jet} can be defined as

$$Q_{\text{jet}} = \sum Q_i \, \eta_i^\kappa \, , \tag{3.66}$$

where Q_i is the charge of the ith particle in the jet and κ is an arbitrary constant. It is evident that the measured jet charge results from $\kappa = 0$. The weight η_i^κ with $\kappa > 0$ is introduced to take account of the fact that particles with larger rapidity have a higher probability to contain the parent quark.

In a recent analysis of data corresponding to 220 pb^{-1} MAC (Ford et al., 1986) have presented the differential cross section for hadronic events using the event thrust axis as the estimator for the quark-antiquark direction in the final state. They have used a value of 0.2 for κ in the above expression for the jet charge and have admitted only events with opposite jet charge. The expected charge asymmetry from electroweak interference at $\sqrt{s} = 29$ GeV, using the weighting algorithm (3.66), is $A_q = (+2.2 \pm 0.5)$ % where the error results from the limited Monte Carlo statistics. The fully corrected data are shown in Fig. 3.43. The charge asymmetry determined from the data amounts to $A_q = (+2.8 \pm 0.5)$ %.

Assuming universality of the different quark flavours, one may derive the average quark axial-vector charge by means of (3.20), taking properly into account threshold factors for the heavy b quark. Assuming furthermore $a_e = -1$, the MAC data yield

$$\langle a_q \rangle = -(1.36 \pm 0.24 \pm 0.20) \, ,$$

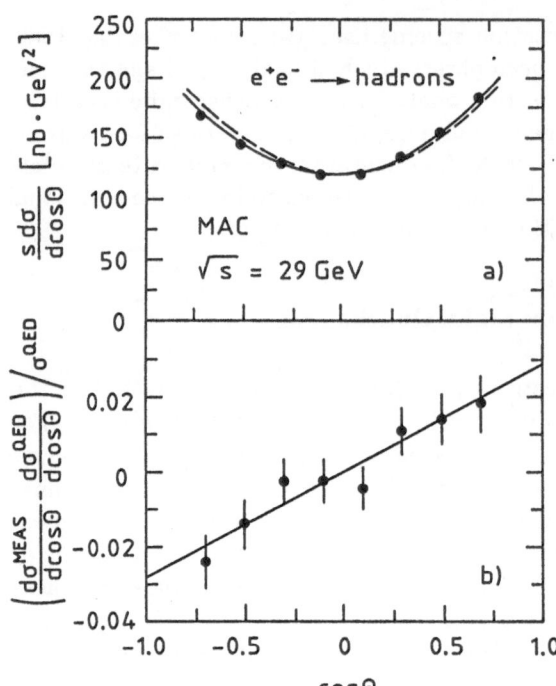

Fig. 3.43. Charge asymmetry for multihadronic two-jet final states, measured by the MAC Collaboration

where the average is taken over down-type quarks and up-type *anti*quarks, as discussed in Sect. 3.1, equation (3.22). The first error given is statistical while the second error is an estimate of the systematics possibly introduced by using a specific value for κ in (3.66) as well as by choosing a specific model for the quark fragmentation. The value above for $\langle a_q \rangle$ is in good agreement with the standard theory expectation of $\langle a_q \rangle = -1$ (see Table 2.1).

A similar analysis has been reported by JADE (Greenshaw et al., 1988), where the jet angular asymmetry was measured at $\sqrt{s} = 35$ and 43 GeV. In their algorithm to define the jet charge, JADE have taken the three fastest particles in each jet of selected two-jet events. Analysing their results for the average quark axial-vector charge, JADE obtain $\langle a_q \rangle = -(1.09 \pm 0.18 \pm 0.23)$, in good agreement with MAC and the standard expectation.

3.6.3 Charge Asymmetries for Heavy Quarks

Much of the information discussed in this section will come from heavy (b) quark final states, where mass effects are expected to contribute. The corrections to the cross section, essentially given by $C_{VV} + C_{AA}$, have already been discussed above (see (3.62)). The antisymmetric term C_A in (3.17), here denoted as C_{VA}^q for the quark flavour $f = q$, receives mass corrections as well, so that the expression for the charge asymmetry for massive final state quarks now reads

$$A_{\mathrm{FB}}^q = \frac{3}{4} \frac{\beta^2 C_{VA}^q}{\frac{\beta}{2}(3 - \beta^2) C_{VV}^q + \beta^3 C_{AA}^q} . \tag{3.67}$$

Besides the QED corrections already discussed for lepton case, one also expects QCD to contribute to the charge asymmetry in the case of quark final states, since the observable is parity-conserving. The QCD corrections have been calculated by Jersak et al. (1981) and yield very small contributions to C_{VA}^q (the corrections to C_{VV}^q and C_{AA}^q have been given in the previous section). The main correction in the lowest order expectation for the charge asymmetry of heavy quarks, however, comes from the β factors in (3.67).

In the framework of the standard theory a sizeable charge asymmetry is expected for $Q\bar{Q}$ final states, which is enhanced with respect to the leptonic case by the charge factor $(e_Q)^{-1}$ (see (3.20)). A measurement of the asymmetry for a given quark requires the determination of both its charge and flavour ("flavour tagging"). Necessarily the algorithms employed will depend on the assumptions about the hadronisation of the quarks to be tagged. Only the heavy quarks charm (c) and bottom (b) provide sufficiently distinct signatures to be separated from the background of the other flavours. Two methods have been employed so far, each one with its characteristic merits and drawbacks.

The first method tries to identify the heavy quark charge and flavour by detecting energetic ("leading") mesons containing the quark under investigation. Charmed quarks may be tagged, e.g., by reconstructing charged D^* mesons from their hadronic decay products in the decay chain

$$
\begin{aligned}
c \to D^{*+} &+ X \\
&\hookrightarrow D^0\,\pi^+ \\
&\qquad\hookrightarrow K^-\,\pi^+ \,.
\end{aligned}
$$

Strong experimental support for the assumption that the D^*'s are indeed fragments of the c quark is provided by the following measurements (see also Bethke, 1985):

— The scaled differential cross section $(s/\beta)d\sigma/dx$ for $e^+e^- \to D^{*\pm}X$, where x is the energy of the D^* in units of the beam energy ($x = 2E_{D^*}/\sqrt{s}$), shows a very similar structure, peaking around $x = 0.7$, over a wide range of energies, from $\sqrt{s} = 10.0$ GeV at CESR (Bebek et al., 1982; Avery et al., 1983) and DORIS II (Albrecht et al., 1985a) and $\sqrt{s} = 29$ GeV at PEP (Yelton et al., 1982; Madaras, 1984; Yamamoto et al., 1985) up to $\sqrt{s} = 34.4$ GeV at PETRA (Althoff et al., 1983; Bartel et al., 1984). Note that the CESR measurements are below the open bottom threshold.

— Assuming equal rates for charged and neutral D^*'s, TASSO (Althoff et al., 1983) find a total D^* cross section of $R(D^*) = 2.5 \pm 1.1$ for $x > 0.3$. This should be compared to the expected inclusive charm and anti-charm cross section of $R = 2 \times 3(2/3)^2 = 2.67$, neglecting small QCD corrections. MARK II (Trilling, 1982), DELCO (Atwood et al., 1983), HRS (Ahlen et al., 1983) arrive at similar conclusions. Background from b quarks cascading through the c quark into D^* can be suppressed by requiring sufficiently energetic D^* mesons, usually selected by their fractional energy x. Although the b quark fragmentation function is hard (Fernandez et al., 1983a; Adeva et al., 1983a; Nelson et al., 1983; Chrin, 1987), producing preferen-

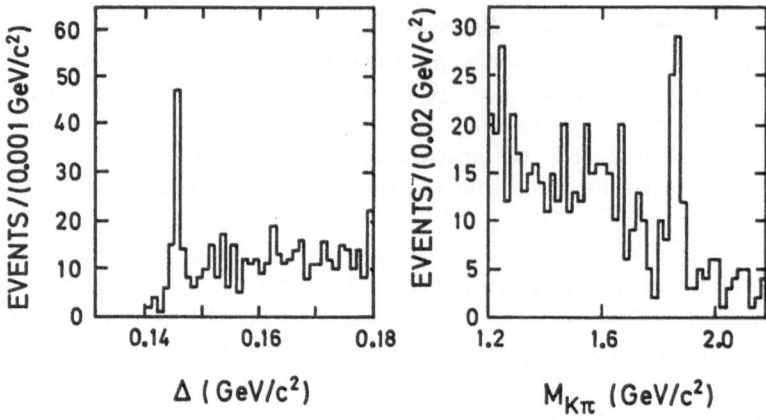

Fig. 3.44. c quark tagging with HRS at PEP (Ahlen et al., 1983a) using D^* and D^0 production. (a) Distribution of $\Delta = M(K\pi\pi) - M(K\pi)$. (b) Distribution of $M(K\pi)$

tially energetic mesons, the fraction of D^* from b decays is estimated to be below 5% for $x > 0.5$ (Suzuki, 1977; Bjorken, 1978).

D^*'s are identified making use of the low Q value (5.7 MeV) involved in the decay $D^* \rightarrow D^0\pi$. Calculating the mass difference $\Delta = M(K\pi\pi) - M(K\pi)$ for $K\pi$ systems in the D region produces a clean signal at $\Delta = 145$ MeV with very little ($\sim 10\%$) background. Identification of K mesons does not seem to be necessary. As an example, the distribution for Δ from HRS (Ahlen et al., 1983a) at PEP is shown in Fig. 3.44. Due to their excellent momentum resolution of $\sigma(p)/p = 0.1\% \, p \, [\text{GeV}]$, HRS can also see the D^0 directly in the $K\pi$ mass spectrum, displayed in Fig. 3.44. Since the branching ratios involved are very small ($B(D^* \rightarrow K\pi\pi) \sim 1.3\%$) statistics is the main problem of the D/D^* tagging method. Experiments have started to include further decays of the D mesons, such as $D^+ \rightarrow K^-\pi^+\pi^+$ and $D^0 \rightarrow K^-\pi^+\pi^+\pi^-$. Still, the total number of tagged events per experiment is only a few hundred, with the exception of HRS, who have obtained almost 700 D, D^* candidates from their complete statistics (Baringer et al., 1988).

Examples of the angular distributions for leading D^* mesons from HRS and JADE are shown in Fig. 3.45 together with fits allowing for a charge asymmetry. The available results on c quark charge asymmetries from the above cited experiments and new results reported at the Munich Conference (see Braunschweig, 1988) are listed in Table 3.12. Also given are the respective expectations from the standard theory using $\sin^2\theta_W = 0.23$ and $M_Z = 92$ GeV. From the asymmetries on obtains the product $a_e \, a_c$ according to (3.20). Assuming $a_e = -1$ leads to the values a_c given in the last column of the table. Although the results are still statistically not very significant, it is worth noting that the asymmetries observed agree in sign and magnitude with the expectations from the standard theory.

The second method to tag heavy quarks, applying to both c- and b quarks, is based on the identification of the charged lepton from the semileptonic decays

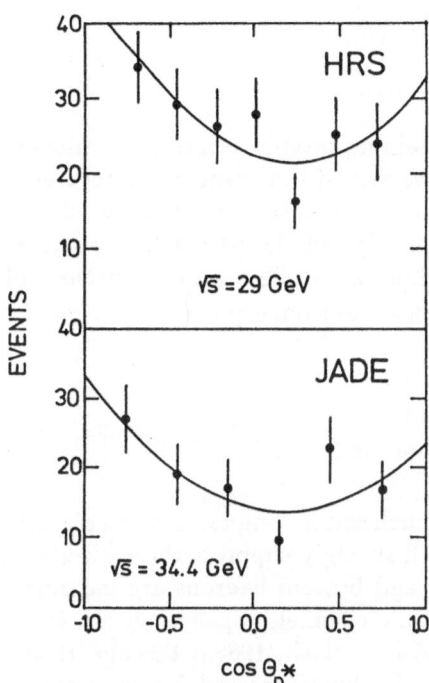

Fig. 3.45. Angular distributions of D^* mesons from HRS and JADE. The full curves result from fits to the data

Table 3.12. Measurements of the charge asymmetries and axial vector coupling constants for c quarks. The value for A_{GSW} is calculated to the same level of corrections as the data, assuming $M_Z = 92$ GeV and $\sin^2\theta_W = 0.23$ for the lowest order standard theory prediction. $a_e = -1$ and $Q_c = +2/3$ is assumed for the calculation of a_c

Exp.	$\sqrt{s}$ [GeV]	Method	A_{meas} [%]	A_{GSW}	a_c
HRS	29	D^*/D	-9.9 ± 2.7	-8.6	$+1.15 \pm 0.32$
MAC	29	μ	-5 ± 11	-3	$+1.6 \pm 3.6$
TPC	29	e, μ	-17.3 ± 10.8	-8.6	$+2.0 \pm 1.3$
TPC	29	D^*	-16.0 ± 16.0	-8.6	$+1.9 \pm 1.9$
CELLO	35.0	e, μ	-8.7 ± 11.4	-13.1	0.65 ± 0.87
CELLO	43.0	e, μ	-16.0 ± 18.4	-20.6	0.70 ± 0.90
JADE	34.4	D^*	-14.0 ± 9.0	-12.6	1.11 ± 0.73
MARK J	35.3	μ	-16.0 ± 9.0	-13.3	1.20 ± 0.69
PLUTO	34.6	μ	-16 ± 18	-12.7	1.24 ± 1.42
TASSO	35.6	D^*	-17.0 ± 9.0	-13.6	1.25 ± 0.68
TASSO	34.6	e	5.0 ± 24.0	-12.7	-0.39 ± 1.87

$$c \rightarrow sl^+ \nu$$

$$b \rightarrow cl^- \bar{\nu} \; .$$

(3.68)

In practice one studies hadronic events containing prompt electrons or muons, i.e. leptons originating from the interaction vertex. If the events are interpreted as semileptonic decays of charm and bottom mesons (or baryons), the charge sign of the lepton is identical with the charge sign of the parent heavy quark. Positive leptons will tag c quarks and b antiquarks, while negative leptons will tag b quarks and c antiquarks. Some confusion arises from b quark cascade decays which lead to "wrong-sign" leptons:

$$b \rightarrow c \quad + \quad \text{hadrons}$$
$$\hookrightarrow l^+ \nu + \text{hadrons} \; .$$

Their spectrum, however, is expected to be much softer compared to the directly produced leptons so that momentum cuts will strongly suppress these events.

The average branching ratios for charm and bottom hadrons are measured at a level of $\sim 10\%$ for each of the semileptonic channels (Spencer et al., 1981; Chadwick et al., 1983; Nelson et al., 1983; Adeva et al., 1983a; Piccolo, 1983). In comparison to D^* tagging, the statistics is thus increased by an order of magnitude. However, serious background problems are encountered when the hadronic events are searched for lepton candidates. The identification of leptons proceeds along the lines discussed above for purely leptonic final states. More restrictive cuts have to be applied, however, since the leptons are accompanied by ordinary hadrons. In the sample of prompt electron candidate events the background arises from pions undergoing a charge exchange reaction or overlapping with energetic photons, or from conversion electrons. The muon sample receives background from decays and punch-through of the numerous hadrons in each event. In addition, purely instrumental background in the muon chambers may lead to accidental associations with non-muonic particles.

The background both for electrons and muons, since originating basically from pions, will show the soft x distribution characteristic for the light quarks. Background reduction and flavour separation is thus attempted by kinematical cuts on the lepton momentum components, exploiting the large mass of the heavy quarks and the fact that the fragmentation functions (at least for b quarks) are hard.

The methods used by the different experiments, although distinctive in details, generally follow the same strategy:

— The transverse momentum p_t of the leptons with respect to the event thrust/sphericity axis is chosen to distinguish between background from light quarks, c quarks and b quarks. Due to the large b quark mass, leptons from b decay will prefer large p_t (≥ 1 GeV/c, e.g.). The hard fragmentation function of the b quarks suggests also restriction to high momentum leptons.

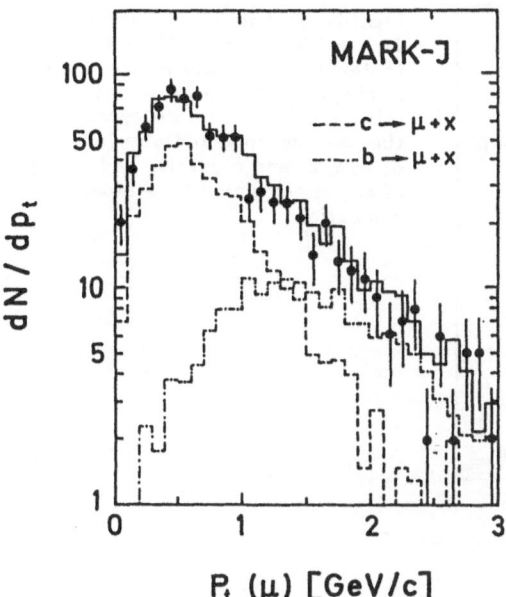

Fig. 3.46. Distribution of the transverse momentum of muons with respect to the thrust axis in multihadronic events from e^+e^- annihilation from the MARK J experiment. The dashed histograms indicate the individual contributions from c- and b quarks. The full histogram is the sum over these and contributions from light quarks

— Events containing heavy quarks can further be enhanced using jet variables and general kinematical cuts: Due to the heavy quark's rest mass the events are expected to have larger jet masses and consequently be more spherical than average. One therefore selects candidate events by imposing conditions on quantities such as thrust, aplanarity, jet masses etc..

One may then either use these variables to select an enriched sample of events containing c or b quarks, or try to fit the variables for the total sample using the semileptonic branching ratios and fragmentation functions as input and/or free parameters. In either case one strongly depends on Monte Carlo calculations to estimate selection efficiencies and the shapes and relative sizes of the contributions from the signal and the various backgrounds. A careful investigation of the systematic errors involved is necessary since the background, particularly for c quarks, is of the same order or even larger than the signal.

As an example the p_t distribution for prompt muons from MARK J (Adeva et al., 1983a) at $\sqrt{s} = 34.5$ GeV is shown in Fig. 3.46. Only muons originating from b quarks seem to be readily separable by a cut in p_t at ~ 1 GeV/c. To extract the signal from c quarks certainly is much harder and does depend stronger on assumptions concerning the competing backgrounds. As a consequence, the parameters of the semileptonic c-quark decay are only poorly known: While most experiments agree on both leptonic branching ratios and fragmentation function for the b quark there are substantial differences for the corresponding quantities of the c quark (Fernandez et al., 1983a; Adeva et al., 1983a; Nelson et al., 1983; Bebek et al., 1982). CELLO (Behrend et al., 1983c) have systematically investigated the semileptonic branching ratios for the c and b quarks and found a significant dependence on the fragmentation function assumed for the charm

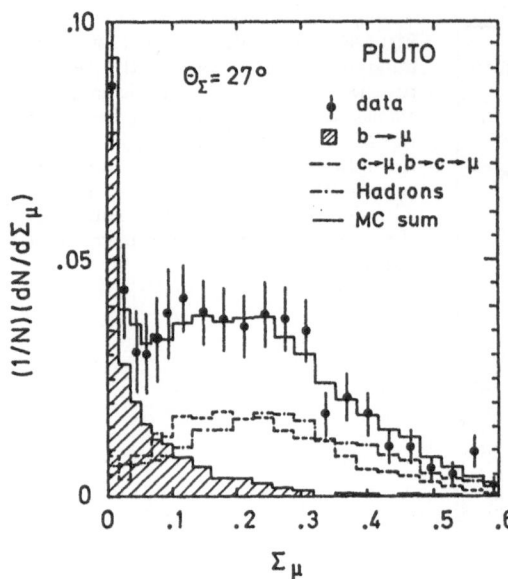

Fig. 3.47. Distribution of the quantity Σ_μ used by the PLUTO collaboration to tag b quarks which have decayed semileptonically into a muon and other particles. The histograms show the various contributions from light quarks, c- and b quarks, and the sum of all, as calculated by a Monte Carlo simulation

quark. No such strong dependence was found for the b quark. Thus the charge asymmetries derived for the b quark will probably be more reliable than those for c quarks. In the following, a few representative analyses will be described, mainly concentrating on the methods for b enhancement.

For the enhancement of b quarks tagged by muons, a single variable Σ_μ has been proposed by the PLUTO collaboration (see Maxeiner, 1985). They describe the isolation in space of the considered lepton by the normalised sum Σ_μ of energies carried by other particles within a cone of opening angle θ_Σ around the lepton. The value for the angle θ_Σ is optimised by Monte-Carlo calculations. Figure 3.47 shows the distribution of the normalised energy sum Σ_μ both for the data and the various expected contributions from heavy and light quarks. A cut in Σ_μ at 0.05, e.g., leads to a b enriched sample with about 80 % b quarks. Thus a rather clean event sample is obtained which suffers, however, of lack of statistics: Only 58 b quark candidates survived the severe cut.

Using the momentum variables p and p_t (with respect to the sphericity axis) for the identified leptons (e and μ), MARK II (Lockyer et al., 1983) have formed three samples: b enhanced, requiring high p and high p_t, c enhanced with high p but lower p_t, and background from the light quarks with both p and p_t low. The combined angular distributions of electrons and muons for the three samples are shown in Fig. 3.48. Also drawn are fits to the data allowing for an asymmetry (full curve) and retaining only the symmetric distribution (dashed line). The b-enhanced sample shows a significant asymmetry, which is, of course, reduced compared to the lowest order GSW expectation due to substantial remaining background from other quarks.

With a very similar analysis CELLO have obtained b (c)-enriched multi-hadronic event samples containing about 750 (1900) electron and muon candidates with typical fractions for the quark signal of about 40 (20) %, respectively,

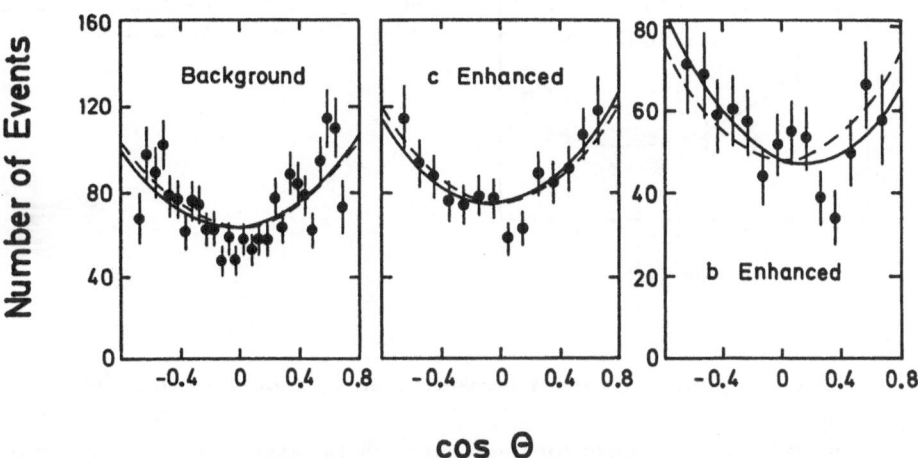

Fig. 3.48. Lepton angular distributions from the MARK II experiment employing cuts to enhance various quark flavours. The full curves show fits to the data. The dashed curves correspond to symmetric distributions

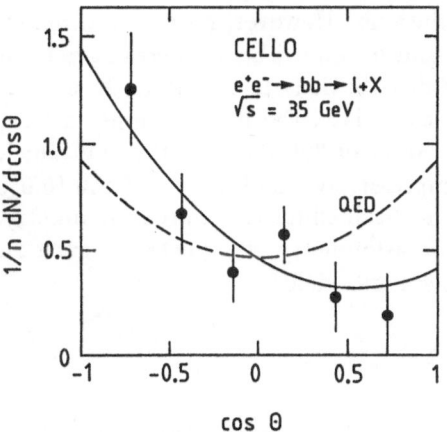

Fig. 3.49. Corrected angular distribution for b quarks tagged via their semileptonic decay from the CELLO experiment at $\sqrt{s} = 35$ GeV. The full curve shows the fit to the data, allowing for an angular asymmetry

in the energy range between 34.5 GeV and the top end of PETRA. As the estimator for the direction of flight of the produced b (c) quarks the event jet axis was used. The background was determined by Monte Carlo methods and subtracted from the raw angular distributions. The differential cross section for $b\bar{b}$ production, summed over both electron and muon semileptonic decays, corrected for radiative and quark mass effects, is shown in Fig. 3.49 together with a fit allowing for an asymmetric angular distribution. A charge asymmetry of $A_b = -(42 \pm 15)\%$ is obtained at $\sqrt{s} = 35$ GeV and $A_b = -(17 \pm 23)\%$ at $\sqrt{s} = 43$ GeV, in resonable agreement with the expectation of $-25.6(-38.6)\%$, respectively, from the standard theory with the usual values for $\sin^2\theta_W$ and M_Z.

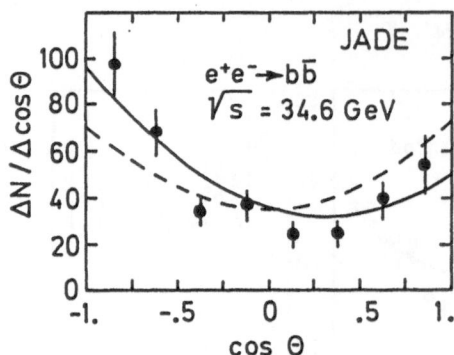

Fig. 3.50. B quark angular distribution from the JADE experiment. The full curve shows the fits to the data. The dashed curves correspond to symmetric QED distribution

CELLO have also given asymmetry measurements for c quarks using the lepton tagging method.

A remarkable measurement for the b quark charge asymmetry has been presented by the JADE collaboration (Bartel et al., 1984a) using the prompt muon tag. In addition to the transverse momentum of the muon, the transverse jet mass and the missing transverse energy carried away by the neutrinos in the semileptonic b decay was studied to distinguish $b\bar{b}$ events from the background. Distributions of these three variables for the various background contributions have been obtained by Monte Carlo calculations. However, no cuts are made, but the fraction of b quarks in the forward and backward hemispheres have been determined in a likelihood fit of the three distributions to the data. In a total sample of 1729 events containing a muon candidate, the fit determines a fraction of b quarks of about 18 %. From the total of 306 b events thus obtained, under a background of 1423 events, the impressively small error of 6.5% (6.0% in a likelihood fit to the asymmetry itself) is obtained for the charge asymmetry. A simple calculation of the error ΔA of the asymmetry A expected in such an analysis of the forward and backward hemispheres yields

$$\Delta A = \frac{1}{f_b} \frac{1}{\sqrt{N}} (1 - A^2) \,,$$

where N is the number of events in the sample (signal and background) and f_b is the fraction of b events in the total sample. Inserting the numbers one would expect a purely statistical error of about 12.5 %. The fully corrected angular distribution for b quarks from the JADE experiment is shown in Fig. 3.50. JADE obtain, at $\sqrt{s} = 34.6$ GeV, a charge asymmetry of -22.8%. The expectation from the standard theory is -23.9%, including quark mass effects which were not corrected for in the analysis.

Table 3.13 summarizes the results on the b quark charge asymmetries. The experiments have usually corrected their angular distributions for background contributions and radiative effects so that in most cases the results can be directly compared to the GSW expectation in lowest order. Also given in the Table are the axial-vector charges of the heavy quarks as deduced from the measured charge

Table 3.13. Measurements of the charge asymmetries and axial-ector coupling constants for b quarks. The value for A_{GSW} is calculated to the same level of corrections as the data, assuming $M_z = 92$ GeV and $\sin^2 \theta_W = 0.23$ for the lowest order standard theory prediction. $a_e = -1$ and $Q_b = -1/3$ is assumed for the calculation of a_b

Exp.	$\sqrt{s}$ [GeV]	Method	A_{meas} [%]	A_{GSW}	a_b
MAC	29	μ	-7 ± 9	-12	-0.6 ± 0.7
MARK II	29	e, μ			$-1.5 \, ^{+0.8}_{-0.7}$
TPC	29	e, μ	-20.4 ± 17.8		-1.15 ± 1.11
CELLO	35.0	e, μ	-42 ± 14.5	-25.6	-1.7 ± 0.75
CELLO	43.0	e, μ	-17.5 ± 23.5	-38.6	-0.42 ± 0.59
JADE	34.6	μ	-22.8 ± 6.5	-25.6	-0.92 ± 0.28
MARK J	37.0	μ	0.0 ± 16.1	-28.8	0.0 ± 0.53
PLUTO	34.8	μ	-36.0 ± 26.2	-25.3	-1.49 ± 1.25
TASSO	34.5	e, μ	-30.0 ± 17.2	-24.8	1.20 ± 0.78

asymmetries. In deriving the axial-vector charges the complete formula (3.67), including purely weak terms, was used: The approximate formula using only the electroweak interference term, which was adequate for the leptons, can be wrong by more than 25 % in the quark case.

Recent measurements of $B^0 - \bar{B}^0$ mixing by the ARGUS collaboration (Albrecht et al., 1987a), however, suggest a further correction to the b quark charge asymmetry measurements: Since the mesons containing a b quark come in three different states depending on the accompanying light quark, i.e. B^+, B^0, and B_s^0, the latter two states may mix with their anti-partners and give rise to "wrong-sign" leptons. The true charge asymmetry is thus reduced depending on the size of the mixing. ARGUS gives a value for the mixing parameter χ_d of neutral, non-strange B mesons of

$$\chi_d = \frac{\Gamma(B^0 \to \bar{B}^0)}{\Gamma(B^0 \to \text{all})} = 0.17 \pm 0.05 \, .$$

With an estimated production ratio of $1 : 1 : 0.3$ for the three quark configurations $b\bar{u}, b\bar{d}$ and $b\bar{s}$, and assuming $\chi_s = 0.25 \pm 0.25$ for the unknown mixing parameter χ_s (bounded within 0 and 0.5) one obtains (see also Wu, 1987)

$$\chi = 0.11 \pm 0.04 \, .$$

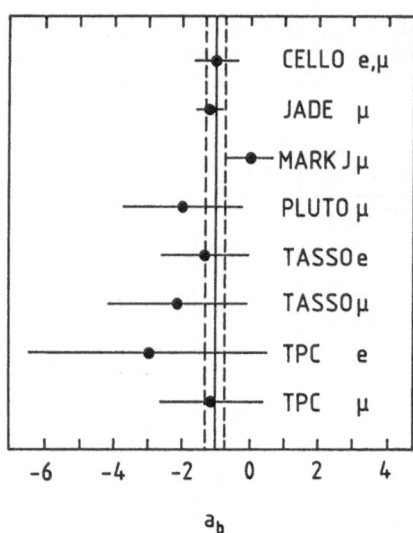

Fig. 3.51. Compilation of measurements for the axial vector coupling constants a_b of the _b_ quark. The measurements are corrected for $B\bar{B}$ mixing (see text)

The measured asymmetry is related to the true one, allowing for mixing, by

$$A_{\text{true}} = \frac{1}{1 - 2\chi} A_{\text{meas}} \, ,$$

where the correction factor is 1.28 ± 0.14. This factor is not included in Table 3.13. A graphic summary of the axial-vector charges, corrected for $B^0 - \bar{B}^0$ mixing, is shown in Fig. 3.51.

Although the errors are still quite large the agreement of the axial-vector charges for _c_- and _b_ quarks with the expectation of the standard theory is surprisingly good. Taking averages from Tables 3.12,13 one finds

$$a_c = 1.12 \pm 0.22$$

and

$$a_b = -1.00 \pm 0.28 \, ,$$

including $B\bar{B}$ mixing, where $+1$ and -1 are expected, respectively.

The axial vector coupling constant a_b is of particular interest since the agreement with the standard theory strongly suggests that the _b_ quark is to be placed in a left-handed doublet with the elusive _t_ quark. Support for the weak isospin doublet assignment comes from a search for _b_ quark decays via flavour-changing neutral currents, i.e.

$$b \to l^+ l^- + X \, , \tag{3.69}$$

which are expected to be strongly suppressed in the standard six quark model with CKM mixing: When the _b_ quark is in an $SU(2)$ singlet one expects a branching ratio at the level of about 1% (Kane and Peskin, 1982). Measurements by CLEO (Kagan et al., 1983), JADE (Bartel et al., 1983a), and MARK J (Adeva et al., 1983b) yield $B(b \to \mu^+\mu^-X) < 0.4\%$ with 95% C.L., definitely below the expectation for a singlet assignment, suggesting that the b quark is a member of an isospin doublet, with its partner still missing. Top has not been found at PETRA and it might not even be in the reach of TRISTAN and LEP I, which will provide centre of mass energies up to 60 and 100 GeV, respectively. Still, as experimental evidence is accumulating, cleverly arranged by the standard theory, we have no choice but to accept that the t quark will be discovered.

An impressive amount of knowledge about the weak couplings of quarks and leptons, in particular of the second and third generations, has been obtained in e^+e^- experiments, strengthening the trust in the correctness of the standard theory in the presently accessible energy range. In the following chapter the available information about the first quark generation will be summarized, which mainly comes from ν scattering and charged lepton hadron scattering and thus is complementary to e^+e^- collisions not only in its physics content but also in its purely experimental aspects.

4. Neutrino Scattering

The scattering of neutrinos on elementary charged fermions has played a key role in discovering the weak neutral current more than a decade ago, providing first evidence for the correctness of the theory of electroweak interactions as formulated by Glashow, Salam and Weinberg. Since that time numerous experiments have been carried out with the aim to provide clean and precise tests of the standard theory. In contrast to the e^+e^- reactions considered in the previous chapter, neutrino scattering proceeds exclusively via the exchange of the intermediate bosons and therefore displays the full effect of the weak neutral current. On the other hand, one has to perform fixed target experiments which are limited in the accessible range for the squared momentum transfer Q^2, typically $Q^2 \leq 100$ GeV$^2/c^2$ for neutrino nucleon scattering with presently available beams.

Counter as well as bubble chamber techniques are employed in neutrino physics. Bubble chambers usually yield clean events with low background, but unfortunately produce low event rates. Electronic experiments, on the other hand, provide high rates due to their large instrumented mass. Here, however, background separation needs careful investigation.

A problem common to all neutrino-induced neutral current experiments is related to the neutrino beams themselves. These beams are derived from decaying π and K mesons which are, for a large number of experiments such as neutrino electron scattering or bubble chamber experiments, not momentum-selected in order to enhance statistics ("wide band" beams). Since the event rates depend strongly on the neutrino energy, the neutrino flux, i.e. energy spectrum and yield, has to be carefully monitored in order to convert the observed numbers of events into cross sections. High energy accelerators provide intense ν_μ and $\bar{\nu}_\mu$ beams with typical average ν energies $\langle E_\nu \rangle \sim 20$ GeV. Low energy (few MeV) $\bar{\nu}_e$ beams can be obtained using nuclear fission reactors. Recently, significant progress has been made with ν_e beams, which are derived from low energy, but high intensity π^+ currents, stopped in a beam dump.

In the subsequent sections, an up-to-date account will be given of the experiments carried out to study neutrino-electron and neutrino-quark (deep inelastic) scattering, some of which have only recently been completed. The discussion is structured according to the various steps towards a unique determination of the couplings of leptons and quarks to the weak neutral current:

- The neutrino electron scattering experiments, in combination with the measurements of the forward-backward asymmetry from e^+e^- annihilation into

μ and τ pairs, pin down the vector and axial-vector couplings v and a of the charged leptons.

— Measurements of the cross section ratios for neutral-current (NC) reactions to charged-current (CC) reactions, scattering neutrinos and antineutrinos on isoscalar targets, allow a precise determination of the sum of the squared chiral couplings for the first generation of quarks, $u_L^2 + d_L^2$ and $u_R^2 + d_R^2$.

— Neutrino scattering on proton and neutron (non-isoscalar) targets provide additional information to separately determine u_L^2, d_L^2, u_R^2 and d_R^2. This is the domain for bubble chamber experiments.

— Exclusive reactions, such as elastic νp, νn scattering and coherent π^0 production allow determining the sign of the individual couplings u_L, d_L, u_R and d_R.

Assuming the standard theory expressions for the chiral couplings, depending on $\sin^2 \theta_W$, the most accurate measurements of the Weinberg angle up to now are derived from neutrino nucleon scattering. The theoretical interpretation of the data in terms of the elementary neutrino quark scattering process depends, however, on assumptions about the structure of the nucleon and effects from the quark-quark interactions inside the nucleon, which cannot be computed a priori. These complications are avoided by considering neutrino electron scattering. While the theoretical interpretation is straightforward, neutrino electron scattering suffers from modest production cross sections due to the small mass of the target electron.

4.1 Experiments

Neutrino interactions have been notoriously difficult to measure owing to their extremely small scattering cross sections which are of order 1 to 1000×10^{-42} cm^2 for neutrino energies around 1 GeV, depending on the mass of the target particle. Therefore very intense neutrino beams and very massive detector systems were mandatory to collect enough statistics for serious tests of the standard theory.

The first νe events have been observed in the heavy liquid bubble chamber Gargamelle at CERN (Hasert et al., 1973), and bubble chambers were instrumental in providing detailed information about the hadronic final states in neutrino nucleon scattering. The more advanced generation of bubble chamber experiments (using the *B*ig *E*uropean *B*ubble *C*hamber (BEBC) at CERN and the 15' chamber at Fermilab), however, made use of additional counter systems, such as *E*xternal *M*uon *I*dentifyers (EMI) and the *I*nternal *P*icket *F*ence (IPF at BEBC), for better event identification and background reduction. The modern high precision neutrino experiments, on the other hand, employ counter techniques exclusively, with massive, highly segmented absorber material, interspersed between active layers for particle detection.

Among the counter experiments one distinguishes two classes according to the density of the absorber material, "light" and "heavy". Detectors using light absorbers are optimised for measuring and discriminating electromagnetic and hadronic showers, since the radiation and interaction lengths of the absorber are of comparable magnitude. Heavy and thick absorbers, on the other hand, will not allow measuring electromagnetic cascades, thus precluding the investigation of neutrino electron scattering. This drawback, however, is counterbalanced by a substantial increase of rate when neutrino hadron reactions are studied.

As an example for the class of light absorbers, the detector of the CERN-Hamburg-Amsterdam-Rome-Moscow (CHARM) collaboration at CERN (Jonker et al., 1982) shall be briefly characterised. It consists of two main components, a fine grain target calorimeter and a muon spectrometer behind it. The target calorimeter, which is surrounded by a frame magnet, is divided into 78 subunits, each one including a marble plate of 300×300 cm^2 surface area and 8 cm thickness, corresponding to about 1 radiation length. The absorber plate is followed by three planes of sensitive elements: A plane of 128 proportional drift tubes with a cross section of 3×3 cm^2 each, a plane of 20 scintillation counters with a cross section of 15×3 cm^2 each, oriented perpendicularly with respect to the direction of the proportional drift tubes, and a plane of 256 limited streamer tubes with a cross section of 1×1 cm^2, oriented parallel to the scintillation counters. The muon spectrometer is made of segmented toroidal iron magnets, interspersed with 18 planes of proportional drift tubes and 6 planes of scintillation counters. The total mass of the detector system is about 750 tons with a fiducial volume for neutrino interactions of about 90 tons.

The CERN-Dortmund-Heidelberg-Saclay (CDHS) Collaboration at CERN (Geweniger et al., 1979) have optimised their detector for neutrino hadron scattering, using iron as absorber material. The detector consists of 21 magnetised iron toroids which are segmented and instrumented with scintillator planes. The assembly combines the functions of target, hadron calorimeter and muon spectrometer. Each toroidal module has a diameter of 3.75 m and an iron thickness of 75 cm. The sampling thickness within a module is 5 cm in the front sections, 15 cm in the middle section, and 75 cm in the last four modules. The scintillator planes in the first 10 modules are made of 15 cm wide strips alternating in the horizontal and vertical direction and viewed by photomultiplier tubes on each side. In the remaining modules the scintillator planes are made of horizontal strips, each 45 cm wide. Between the modules drift chambers are inserted with three coordinate planes for muon tracking. The fiducial mass of the CDHS detector is about 460 tons.

A detector with similar properties as CHARM concerning the ability to measure electrons and muons is run by the Brookhaven-Brown-KEK-Osaka-Pensylvania-Stony Brook (E734) Collaboration at the Brookhaven AGS (Ahrens et al., 1983). At Fermilab, the Caltech-Columbia-Fermilab-Rochester (CCFR) Collaboration (Reutens et al., 1985) uses iron as absorber material, like CDHS, with approximately the same mass.

This short descripton of some major neutrino detector facilities does in no way adequately represent the ingenuity and efforts invested in building beautiful experiments for this rich field of research. Further references to the other detectors not mentionned here will be given below together with the presentation of their physics results.

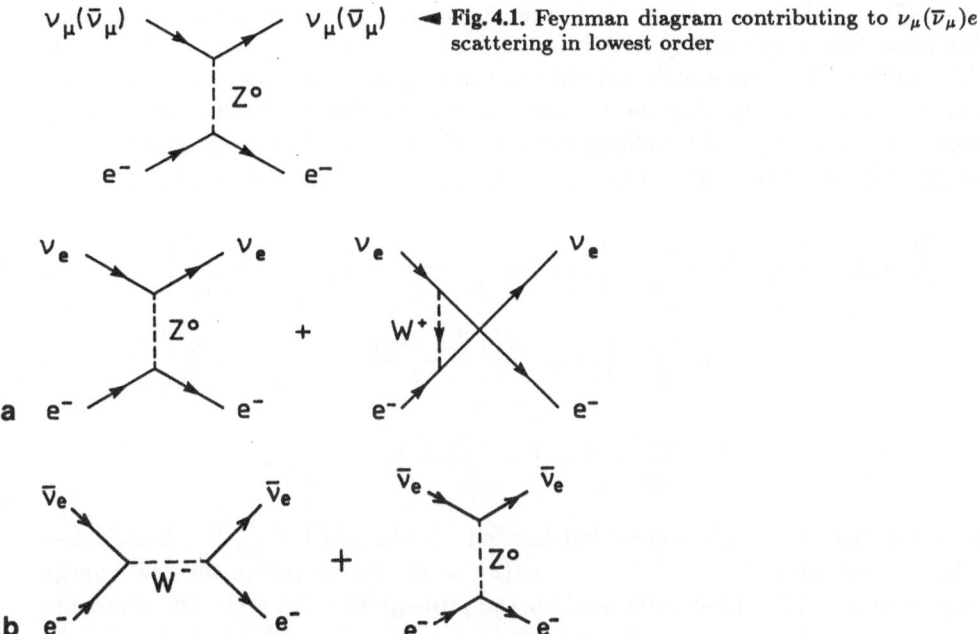

Fig. 4.2. Feynman diagrams contributing to $\nu_e e$ (a) and $\bar\nu_e e$ (b) scattering in lowest order

4.2 Neutrino-Electron Scattering

Neutrino-electron elastic scattering is of particular theoretical interest since only free, pointlike particles are involved. The corresponding cross sections can be calculated with high precision, including radiative corrections, using only the framework of the standard theory. There are four νe reactions accessible by experiment with present-day neutrino beams, which are mediated by the neutral current (NC), i.e. by Z^0 exchange:

$$\nu_\mu\, e^- \rightarrow \nu_\mu\, e^- \tag{4.1}$$

$$\bar\nu_\mu\, e^- \rightarrow \bar\nu_\mu\, e^- \tag{4.2}$$

$$\bar\nu_e\, e^- \rightarrow \bar\nu_e\, e^- \tag{4.3}$$

$$\nu_e\, e^- \rightarrow \nu_e\, e^-. \tag{4.4}$$

The first two reactions proceed only via Z^0 exchange in the t channel according to the diagrams in Fig. 4.1. For the electron neutrino induced reactions (4.3, 4) one also expects the charged current (CC), i.e. $W^\pm$ exchange, to contribute. The relevant lowest order Feynman diagrams for these two reactions are shown in Fig. 4.2 .

Differential cross sections for the reactions (4.1, 2) can be formally obtained from the t-channel helicity amplitudes for $e^+e^- \to e^+e^-$, given in Table 3.6. To establish the correspondence, only the weak part of the coefficients $\varepsilon_{ij}(w)$ is effective, and $i(j)$ denotes the handedness for the electron (neutrino). Taking into account both possible helicity states for the electron by the appropriate spin weight $1/2$, the differential cross section for $\nu_\mu e^- \to \nu_\mu e^-$, e.g., thus reads:

$$\frac{d\sigma}{d\Omega}(\nu_\mu e^- \to \nu_\mu e^-) = \frac{\alpha^2 s}{4} \cdot \frac{1}{2} \left[\left| \frac{1}{t} \varepsilon_{\mathrm{RL}} \left(\frac{1 + \cos\theta}{2} \right) \right|^2 + \left| \frac{1}{t} \varepsilon_{\mathrm{LL}} \right|^2 \right]$$

$$= \frac{s}{32\pi^2} \left[\left| 8 \, c_{\mathrm{R}} \, \frac{\varrho G}{\sqrt{2}} \left(\frac{1 + \cos\theta}{2} \right) \right|^2 + \left| 8 \, c_{\mathrm{L}} \, \frac{\varrho G}{\sqrt{2}} \right|^2 \right] \quad (4.5)$$

$$= \frac{s \varrho^2 G^2}{4\pi^2} \left[c_{\mathrm{R}}^2 \left(\frac{1 + \cos\theta}{2} \right)^2 + c_{\mathrm{L}}^2 \right].$$

In these expressions the purely left-handed coupling of the ν_μ has been taken into account by setting $c_{\mathrm{L}}(\nu) = 1$, $c_{\mathrm{R}}(\nu) = 0$. Furthermore the low energy approximation (3.11) has been used for the propagator term $g(t)$. The constants c_{R} and c_{L} are the familiar right- and left-handed couplings of the electron to the weak neutral current.

It is customary in ν interactions to work in the laboratory system (here: electron rest system) and to introduce the variable $y = T_e/E_\nu$, which is the ratio of the kinetic energy of the electron in the laboratory to the incoming neutrino energy. Transforming to the laboratory system, the angular dependence $(1 + \cos\theta)/2$ is replaced by $(1 - y)$. One obtains

$$\frac{d\sigma}{dy}(\nu_\mu e \to \nu_\mu e) = \frac{2\varrho^2 G^2 m_e E_\nu}{\pi} \left[c_{\mathrm{L}}^2 + c_{\mathrm{R}}^2 (1 - y)^2 \right], \quad (4.6)$$

and analogously for antineutrino electron scattering

$$\frac{d\sigma}{dy}(\bar\nu_\mu e \to \bar\nu_\mu e) = \frac{2\varrho^2 G^2 m_e E_\nu}{\pi} \left[c_{\mathrm{R}}^2 + c_{\mathrm{L}}^2 (1 - y)^2 \right]. \quad (4.7)$$

Total cross sections are obtained by integration over y, which yields the well-known factor $1/3$ in front of the term coupling the neutrino to the opposite-helicity electron. Expressing the chiral couplings by the vector and axial-vector constants v_e and a_e (see (2.36)) leads to the following formulae for the total cross sections:

$$\frac{1}{E_\nu}\sigma(\nu_\mu e \to \nu_\mu e) = \frac{\varrho^2 G^2 m_e}{8\pi} \left[(v + a)^2 + \frac{1}{3}(v - a)^2 \right]$$

$$\frac{1}{E_{\bar\nu}}\sigma(\bar\nu_\mu e \to \bar\nu_\mu e) = \frac{\varrho^2 G^2 m_e}{8\pi} \left[(v - a)^2 + \frac{1}{3}(v + a)^2 \right]. \quad (4.8)$$

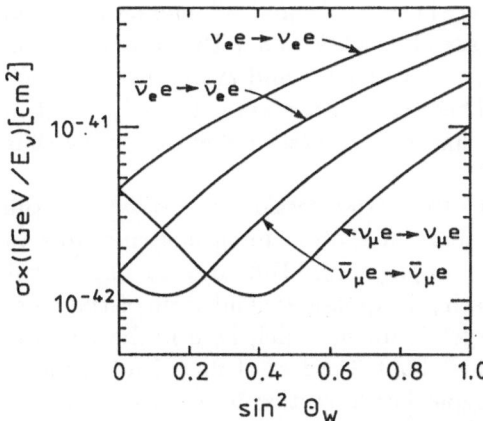

Fig. 4.3. Cross sections for νe scattering in lowest order as function of $\sin^2 \theta_W$

Note that due to the form of the above equations, measurements of the total cross sections will limit the coupling constants v and a to lie within elliptical rings, centred at $v = 0$, $a = 0$. The cross sections are shown in Fig. 4.3 together with those for electron-neutrino induced reactions which are discussed later. Using the standard expressions for v and a in Table 2.1, the cross sections depend on a single common parameter, $\sin^2 \theta_W$. The dependence on $\sin^2 \theta_W$ is also shown in Fig. 4.3.

When comparing these lowest order predictions with experiment, radiative corrections should be taken into account. In neutrino scattering it has become customary to introduce the radiative corrections into the Born term expressions, rather than to correct the data to the Born level, as is usually done in e^+e^- physics. These corrections depend, as for the case of e^+e^- scattering presented in Sect. 3.2, on the renormalisation scheme chosen. In neutrino scattering one usually works in the on-shell scheme which uses the Fermi coupling constant G, the fine structure constant α and the physical masses of the weak bosons (compare to (3.10) in the case of e^+e^- scattering). $\sin^2 \theta_W$, the only free parameter in the above cross section formulae, is defined in terms of the weak boson masses through (2.16):

$$\sin^2 \theta_W = 1 - \frac{M_W^2}{M_Z^2} .$$

Radiative corrections for neutrino electron scattering have been discussed by various authors (see, e.g. Sirlin, 1980; Marciano and Sirlin, 1980, and, in particular, Sarantakos et al., 1983). Their effect is to modify v_e and a_e in the following way:

$$v_e = \rho(-1 + 4\kappa \sin^2 \theta_W) \qquad a_e = \rho(-1) , \qquad (4.9)$$

with two new factors ρ and κ, which are equal to 1 at the Born level. They assume values slightly different from 1 when higher order radiative corrections are included. As usual, the electroweak radiative corrections depend on the fermion masses, on the masses of the heavy bosons $W^{\pm}$ and Z^0, and on the mass of the Higgs scalar. Assuming a top mass of 45 GeV and a Higgs mass of 100 GeV one obtains the corrections $\Delta\varrho = \rho - 1 = +0.005$ and $\Delta\sin^2\theta_W = -0.002$ for a central value of $\sin^2\theta_W = 0.23$. These corrections are extremely small in comparison to the present experimental uncertainties (see below) and may thus be safely neglected.

In the above expressions (4.8) for the total cross sections one recognises the well-known linear dependence on the energy of the incoming neutrino, due to the low energy limit for the W and Z propagators. But also an unpleasant experimental problem in neutrino scattering is apparent: The scale of the cross section is given by the mass of the target fermion, which here is the electron mass. One thus expects νe cross sections about 3 orders of magnitude smaller than in neutrino nucleon scattering. A typical cross section in νe scattering is of order 10^{-42} cm^2 or 10^{-9} nb. Since the kinematics in νe scattering is well defined by measuring only the outgoing electron, one is able to make use of wide band ν beams, for which no beam energy information is available on an event-to-event basis but which provide the largest ν fluxes and thus the highest νe event rates.

The experimental difficulties in extracting elastic νe events from the data are quite diverse. First of all there is the overwhelming background from $\nu\,(\bar{\nu})$ nucleon neutral and charged current reactions. This background can largely be removed by exploiting the marked difference in the shower development of a single electron as opposed to (many) hadrons. Charged current events with identified muons are trivially separated. The remaining events are mostly due to $\nu_e\,(\bar{\nu}_e)$ induced quasi-elastic charged current events (e.g. $\bar{\nu}_e\,p \to e^+ n$) and neutral current events with photons from a π^0 in the final state, produced by coherent scattering of $\nu\,(\bar{\nu})$ on nuclei. While the quasi-elastic CC events produce electrons at substantial scattering angles and can thus be subtracted from the νe sample (see below), a certain fraction of events from coherent scattering are recognised by studying the difference in the shower development of photons and electrons in the first radiation length. Fine grain calorimeters are mandatory for such detailed shower studies.

The two-body kinematics of νe elastic scattering is simple, and is completely defined by the only particle observable in the reaction, i.e. the scattered electron. Consider the general expression for the momentum transfer Q^2 from the initial neutrino with energy E_ν to the final electron with energy E and scattering angle θ with respect to the (known) neutrino direction:

$$Q^2 = 4\,E_\nu\,E\,\sin^2\left(\frac{\theta}{2}\right).$$

An equivalent relation for Q^2 can be written down using the target electron of mass m and the scattered neutrino, expressed in terms of the scattered electron:

$$Q^2 = 2\,m\,(\,E_\nu - E\,).$$

Comparing these two relations one obtains, approximating $\sin(\theta/2)$ by $\theta/2$ as

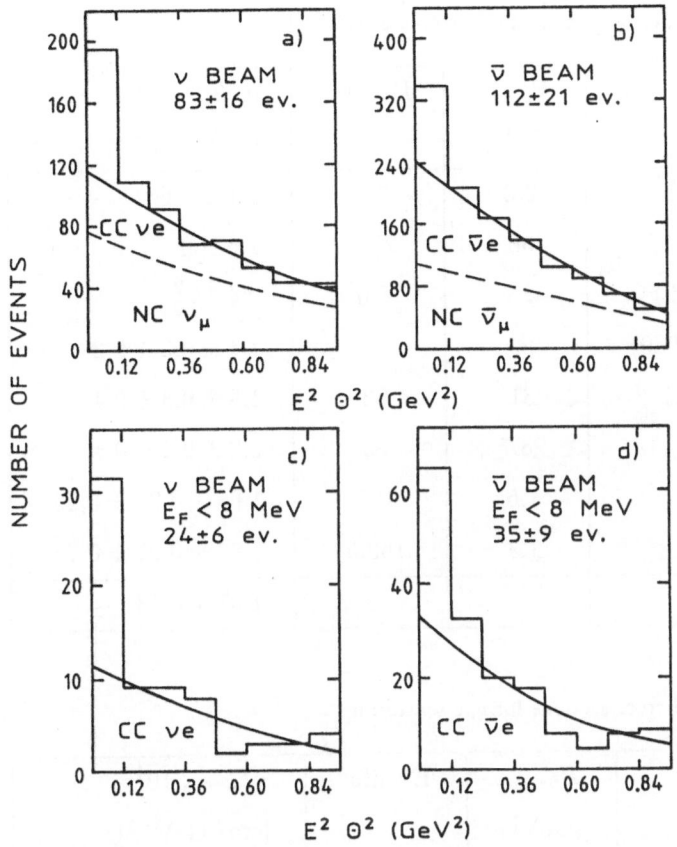

Fig. 4.4. Distributions of the quantity $E^2\theta^2$ for νe scattering candidates from the CHARM experiment

justified by the small scattering angles expected

$$E\,\theta^2 = 2\,m\left(1 - \frac{E}{E_\nu}\right).\qquad(4.10)$$

Plotting the variable $E\,\theta^2$ (or $E^2\,\theta^2$, as used by some experiments for historical reasons) will show a sharp peak at small values due to the factor m in (4.10), superimposed on the residual, essentially flat background from inelastic processes. As an example, Fig. 4.4 shows the distribution of $E^2\,\theta^2$ from the CHARM collaboration (Bergsma et al., 1984). A clear signal above the coherent scattering and CC background is observed. For the determination of the νe cross sections the neutrino flux is monitored by quasi-elastic CC reactions such as $\nu_\mu\,n \to \mu^-\,p$ and $\bar\nu_\mu\,p \to \mu^+\,n$, where backgrounds due to additional pion production can be controlled by analysing the shower activity close to the interaction vertex.

A compilation of $\nu_\mu\,e$ and $\bar\nu_\mu\,e$ cross sections measurements from CHARM and other experiments, normalised to the incident neutrino beam energy, is pre-

Table 4.1. Compilation of cross sections for $\nu_\mu e$ scattering

Reference	E_{beam} [GeV]	Events	$\sigma/E_\nu \times 10^{42}$ [cm^2 GeV^{-1}]
Faissner et al. (1978)	2.2	7.1	1.1 ± 0.6
Cnops et al. (1978)	30	20.5	1.6 ± 0.4
Armenise et al. (1979)	25	8.6	$2.4 {}^{+1.2}_{-0.9}$
Heisterberg et al. (1980)	20	34	$1.4 \pm 0.3 \pm 0.2$
Bergsma et al. (1984)	31	83	$1.8 \pm 0.3 \pm 0.4$
Zacek (1988)	28.7	83	$2.2 \pm 0.4 \pm 0.4$
Ahrens et al. (1985)	1.5	107	$1.60 \pm 0.29 \pm 0.26$
Diwan (1988)	1.3	159.5	$1.87 \pm 0.21 \pm 0.26$
mean			1.67 ± 0.18

Table 4.2. Compilation of cross sections for $\bar{\nu}_\mu e$ scattering

Reference	E_{beam} [GeV]	Events	$\sigma/E_\nu \times 10^{42}$ [cm^2 GeV^{-1}]
Faissner et al. (1978)	2.0	6.3	2.2 ± 1.0
Bergsma et al. (1984)	24	112	$1.5 \pm 0.3 \pm 0.4$
Zacek (1988)	23	112	$1.6 \pm 0.3 \pm 0.3$
Ahrens et al. (1985)	1.4	59	$1.16 \pm 0.20 \pm 0.14$
Diwan (1988)	1.4	99.7	$1.20 \pm 0.17 \pm 0.13$
mean			1.31 ± 0.19

sented in Tables 4.1,2, respectively, where early experiments leading only to upper limits (Blietschau et al., 1978; Astratyan et al., 1979; Pullia et al., 1979; Armenise et al., 1979; Berge et al., 1979) have been omitted. The table also contains recent updates by both CHARM (Zacek, 1988) and E734 (Diwan, 1988). The units for the cross sections are chosen as 10^{-42} [cm^2 GeV^{-1}]. This numerical constant is very close to the coefficents in front of the brackets in (4.8), which is

$$\varrho^2 G^2 m_e/8\pi = 1.077 \times 10^{-42} [\text{cm}^2\,\text{GeV}^{-1}]$$

when $\varrho = 1$, as suggested by standard theory assuming a minimal Higgs sector. The mean values for the normalised cross sections shown in the tables have been calculated by weighting the individual measurements with their statistical and sytematic errors added in quadrature. For CHARM and E734 only the 1988 updates have been used.

Before trying to extract the value of $\sin^2 \theta_W$ from the neutrino electron scattering data, it is instructive to concentrate on the more general analysis of the vector and axial-vector charges v and a of the electron. From equations (4.8) it is clear that each of the νe cross sections will constrain the allowed values of v and a to lie within an elliptical ring in the $v - a$ plane, centered at the origin (see Fig. 4.5 below). Taking sums and differences of νe and $\bar{\nu}_\mu$ cross sections one can isolate the squares of the coupling constants v, a and their product:

$$\frac{1}{E_\nu}\left[\sigma(\nu_\mu e) + \sigma(\bar{\nu}_\mu e)\right] = \frac{\varrho^2 G^2 m_e}{8\pi}\frac{8}{3}(v^2 + a^2)$$

$$\frac{1}{E_\nu}\left[\sigma(\nu_\mu e) - \sigma(\bar{\nu}_\mu e)\right] = \frac{\varrho^2 G^2 m_e}{8\pi}\frac{8}{3}v\,a\;.$$

(4.11)

Inserting the weighted averages from Table 4.1 into (4.11) yields, with the assumption $\varrho = 1$, the following relations:

$$v^2 + a^2 = 1.04 \pm 0.13$$

$$va = 0.13 \pm 0.14\;.$$

(4.12)

The product va is compatible with zero, so that either v^2 or a^2 is close to 1. This is precisely the expectation from the standard theory with $\sin^2 \theta_W$ in the vicinity of $1/4$, i.e. $v^2 \sim 0$, $a^2 = 1$.

The four-fold ambiguity in (4.12) can be reduced by considering $\bar{\nu}_e e$ scattering, the cross section of which was given in (4.3). In a reactor experiment, Reines et al. (1976) have measured the cross section for elastic anti-neutrino electron scattering for two regions of the outgoing electron energy, 1.5 MeV $\leq E_e \leq$ 3 MeV and 3 MeV $\leq E_e \leq$ 4.5 MeV, corresponding to two energy regions of the incident $\bar{\nu}_e$. The neutrino energies in such a reactor are so low (few MeV), that the mass of the electron can no longer be neglected, as was done for the accelerator experiments (see (4.8)). Furthermore the s-channel W exchange has to be included for the theoretical prediction of the cross section, as indicated in Fig. 4.2. The cross section for $\bar{\nu}_e e \to \bar{\nu}_e e$ is given (see, e.g., t'Hooft, 1971; Chen and Lee, 1972; Kim, 1977) by

$$\frac{d\sigma}{dy}(\bar{\nu}_e\, e \to \bar{\nu}_e\, e) = \frac{2\varrho^2\, G^2\, m_e E_{\bar{\nu}}}{\pi}\left[c_R^2 + (1/\varrho + c_L)^2\,(1 - y)^2\right.$$

$$\left. - \frac{m_e}{E_{\bar{\nu}}}\,y\,(1/\varrho + c_L)\,c_R\right]\;.$$

(4.13)

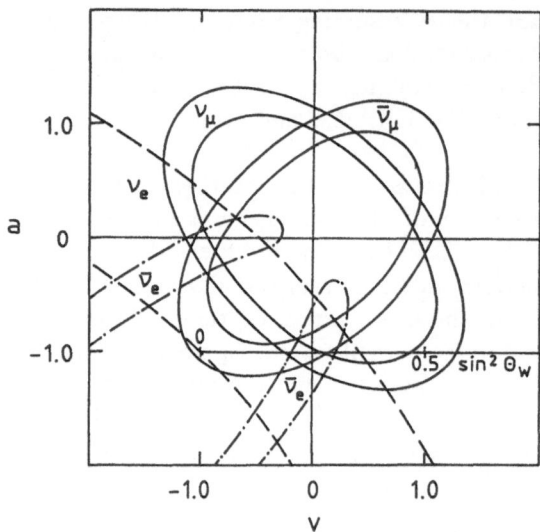

Fig. 4.5. 68% C.L. contours in the $v - a$ plane from the available νe scattering data

The charged current contribution is recognised in the terms proportional to $1/\varrho$, ϱ being present due to the a priori unknown relative strength between charged and neutral currents. Integrating over y and substituting for v and a one obtains

$$\frac{1}{E_{\bar{\nu}}} \sigma\left(\bar{\nu}_e\, e \to \bar{\nu}_e\, e\right) = \frac{\varrho^2 G^2 m_e}{8\pi} \left[(v - a)^2 + \frac{1}{3}\left(4/\varrho + v + a\right)^2 \right.$$
$$\left. - \frac{m_e}{2E_{\bar{\nu}}}\left(4/\varrho + v + a\right)\left(v - a\right)\right]. \tag{4.14}$$

Similar to the muon-neutrino cross sections (4.8), $\bar{\nu}_e e$ scattering constrains the values of v and a to lie within elliptical rings. In contrast to $\nu_\mu e$, however, the rings are centred at $(v = -2/\varrho,\ a = -2/\varrho)$, as can be seen from (4.14). Therefore $\bar{\nu}_e e$ scattering is sensitive to the absolute sign of v and a, which is due to the interference of the neutral with the charged current.

In a thorough analysis of the weak neutral couplings of the electron, Krenz (1985) has combined the available data from reactions (4.1) and (4.2) (see Tables 4.1, 2) with the measurements on $\bar{\nu}_e e$ scattering in the two quoted energy regions. Due to the strong energy dependence of the neutrino cross sections, very careful studies of the $\bar{\nu}_e$ spectrum from the reactor are mandatory. For the appropriate weighting of the neutrino spectrum within a given energy region, correction factors are introduced into (4.14), which multiply the various terms in the neutrino electron cross section (see Avignone and Greenwood, 1977). Using their results, Krenz has determined the allowed regions in the $v - a$ plane for the three neutrino electron reactions discussed so far, assuming $\varrho = 1$. The 68% C.L. contours extracted from the individual cross sections for all three reactions are shown in Fig. 4.5. The contours for $\bar{\nu}_e e$ scattering differ somewhat from earlier results by Kim et al. (1981), who based their work on a parametrisation of the energy-weighted neutrino cross section as given by Liede et al. (1978).

As is evident from Fig. 4.5, the fourfold ambiguity is resolved, leaving two allowed regions for the electron couplings to the weak neutral current. Fitting for a common set of v, a values for the three v reactions, Krenz (1985) obtains an axial vector-like solution

$$v_e = -0.076 \pm 0.094 \qquad a_e = -0.990 \pm 0.052$$

and a vector-like solution

$$v_e = -0.984 \pm 0.052 \qquad a_e = -0.102 \pm 0.090 \ .$$

The corresponding 95% C.L. contours combining all ve data have already been shown in Fig. 3.36, when e^+e^- data were discussed. Assuming universal coupling of the weak neutral currents, the vector-like solution is excluded by e^+e^- data. The remaining axial vector-like solution is in agreement with the standard theory, the prediction of which for different values of the Weinberg angle is also shown in the v, a plane of Fig. 3.36.

In a next step one may use the standard theory relations from Table 2.1 to express v and a by the single parameter $\sin^2\theta_W$. A fit for the weak mixing angle as the only free parameter yields (Krenz, 1985)

$$\sin^2\theta_W = 0.231 \pm 0.023 \ .$$

This result has been obtained by again assuming $\varrho = 1$. If ϱ is left as a free parameter the fit results are

$$\sin^2\theta_W = 0.231 \pm 0.024$$

$$\varrho = 0.989 \pm 0.052 \ .$$

Similar to the analysis of the e^+e^- data presented in Sect. 3.5, one can dispose of the ϱ dependence by using expression (3.9) for the propagator term $g(s)$. The free parameters describing the scattering process are then $\sin^2\theta_W$ and M_Z, with the increased sensitivity to $\sin^2\theta_W$ due to the propagator parametrisation (3.9) already encountered in e^+e^- scattering (see Chap. 3). Constraining the mass of the Z^0 to the recent experimental values 92 GeV (see Chap. 6) yields the value

$$\sin^2\theta_W = 0.216 \pm 0.014 \ .$$

While the above analyses rely on theoretical (assuming $\varrho = 1$) or experimental (mass of the Z^0) input to determine the Weinberg angle, one can eliminate such uncertainties by forming the ratio of neutrino to anti-neutrino scattering cross sections

$$R = \sigma\left(v_\mu e \rightarrow v_\mu e\right)/\sigma\left(\bar{v}_\mu e^- \rightarrow \bar{v}_\mu e^-\right) \ .$$

In the ratio the relative coupling strength parameter ϱ has disappeared and the physical observables depends now only on the Weinberg angle. Furthermore, on purely experimental grounds, the cross section ratio has the additional advantage

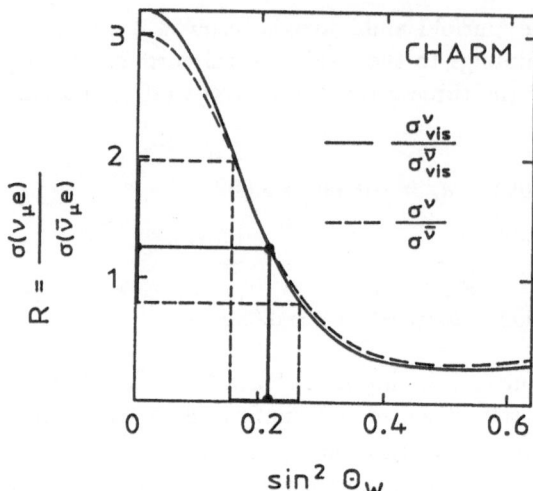

Fig. 4.6. Determination of the Weinberg angle from the measurement of $\nu_\mu e$ scattering by the CHARM experiment

to minimise systematic errors, arising, e.g. , from efficiency calculations and the absolute normalisation of the neutrino flux for the cross section determination. Such errors tend to cancel in the cross section ratio. Using the cross sections in (4.8) and v, a from Table 2.1 one obtains

$$R = 3 \, \frac{1 - 4 \sin^2 \theta_W + (16/3) \sin^4 \theta_W}{1 - 4 \sin^2 \theta_W + 16 \sin^4 \theta_W} \, . \tag{4.15}$$

Since R only depends on $\sin^2 \theta_W$, it has no theoretical uncertainty apart from minor uncertainties in the calculation of the radiative corrections due to some unknown masses as mentioned above. It is thus, in principle, the best estimator for the Weinberg angle besides the mass ratio of the weak bosons. To indicate the sensitivity of the cross section ratio to the Weinberg angle, R is shown in Fig. 4.6 as a function of $\sin^2 \theta_W$ according to the theoretical expectation (4.15), and determined under real experimental conditions (solid line, taken from Bergsma et al., 1984). From the figure one recognises the enhanced sensitivity of R for values of $\sin^2 \theta_W$ close to $1/4$, where the error on $\sin^2 \theta_W$ is given by

$$\delta \sin^2 \theta_W = \frac{1}{8} \frac{\delta R}{R} \, .$$

Both the CHARM (Bergsma et al., 1984) and E734 (Ahrens et al., 1985) collaborations have measured R. Their results on R, based on the new measurements quoted in Tables 4.1,2, and the values for $\sin^2 \theta_W$ extracted by means of (4.15) are given in Table 4.3. Both experiments agree well, and combining these determinations, which entirely dominate the available data in precision, yields the following world average for $\sin^2 \theta_W$ from neutrino electron scattering

$$\sin^2 \theta_W = 0.201 \pm 0.021 \, .$$

Table 4.3. Measurements of the cross section ratio $R = \sigma(\nu_\mu e)/\sigma(\bar\nu_\mu e)$ and resulting values for $\sin^2\theta_W$

Experiment	R	$\sin^2\theta_W$
E734	$1.56^{+0.31+0.18}_{-0.25-0.17}$	$0.197^{+0.020\ +0.014}_{-0.021\ -0.013}$
CHARM	1.20 ± 0.35	$0.211 \pm 0.035 \pm 0.011$

As mentioned earlier in this section, statistics is limiting the accuracy in the determination of the Weinberg angle using electron neutrino scattering. In an attempt to significantly improve the error on $\sin^2\theta_W$, the Charm Collaboration has built a new detector, CHARM II, 690 t of mass, using glass as target material, and aiming at a precision of 0.005 in $\sin^2\theta_W$. This demands a more than tenfold increase in event statistics as compared to the above experiments. The detector has started taking data in 1987, collecting about 300 events in either neutrino electron channel. A preliminary determination of the Weinberg angle from these data shows agreement with the above world average (see Zacek, 1988).

To complete the subject of neutrino electron scattering, two other neutrino electron reactions should be mentioned. The first one is the reaction $\nu_\mu e^- \rightarrow \nu_e \mu^-$, or inverse μ decay. This reaction is mediated by the charged current only and will therefore not provide any information on weak neutral currents. It is thus not further discussed here, although measurements exist (Jonker et al., 1980).

Of some interest for the standard theory, on the other hand, is the reaction $\nu_e e \rightarrow \nu_e e$ which has recently been measured for the first time (Allen et al., 1985). This reaction is mediated by both the neutral and the charged current (see Fig. 4.1). The corresponding cross section is given by

$$\frac{d\sigma}{dy}(\nu_e\, e \rightarrow \nu_e\, e) = \frac{2\varrho^2\, G^2\, m_e\, E_\nu}{\pi} \left[c_R^2\,(1-y)^2 \right.$$
$$\left. + (1/\varrho + c_L)^2 - \frac{m_e}{E_\nu}\, y\,(1/\varrho + c_L)\, c_R \right].$$

After integration and changing to the vector and axial-vector charges one obtains

$$\frac{1}{E}\,\sigma\,(\nu_e\, e \rightarrow \nu_e\, e) = \frac{\varrho^2 G^2 m_e}{8\pi} \left[(4/\varrho + v + a)^2 + \frac{1}{3}\,(v-a)^2 \right.$$
$$\left. - \frac{m_e}{2E_\nu}\,(4/\varrho + v + a)\,(v - a) \right].$$

(4.16)

From the dependence of the cross section on v and a one can see that this reaction will not help in resolving the twofold ambiguity of the neutrino scattering data. One expects, as in the case of $\bar\nu_e\, e$ scattering, a confidence interval of the shape of an elliptical ring, centered at $v = -2/\varrho$, $a = -2/\varrho$. However, as pointed out,

117

e.g., by Kayser et al. (1979), one can test the sign of the interference between the amplitudes from the $W^{\pm}$ and Z^0 exchanges. Such an interference is also expected for $\bar{\nu}_e e$ scattering in the reactor experiments (see (4.14)). There, however, the experimental results were not conclusive.

In the standard theory, the interference term for $W^{\pm}$ and Z^0 exchange in $\nu_e e$ scattering is predicted to be destructive, as can be seen from (4.16): Neglecting the third term multiplied by m_e, the part of the cross section due to the $W - Z$ interference is proportional to $+(v + a)$. Since a is equal to -1 and $v \sim 0$, the interference term is negative.

The experiment to measure $\nu_e e$ scattering is carried out at the Los Alamos National Laboratory (LAMPF) where an intense source of ν_e is derived from the 600 μA proton beam of 765 MeV kinetic energy, producing positive pions which are stopped in a beam dump. The stopped π^+'s decay in the beam dump, leading to ν_μ's, followed by stopped μ^+ decays leading to $\bar{\nu}_\mu$'s and ν_e's. The average energy of the final electron neutrinos is around 25 MeV. Since the strength of the charged current interaction is appreciably larger than the neutral current interaction for neutrino electron interactions at $\sin^2\theta_W \sim 0.25$ (see also Fig. 4.3), one expects the $\nu_e e$ scattering to dominate $\nu_\mu e$ and $\bar{\nu}_\mu e$ scattering.

Due to the small expected event rate the detector was designed to effectively reduce cosmic ray background by means of three active and passive shields of steel and concrete with effective thickness of over 900 g/cm^2, surrounding a fine-grained sandwich detector of plastic scintillator and flash chamber modules. The observed number of $\nu_e e$ scattering events after subtraction of $\nu_\mu e$, $\bar{\nu}_\mu e$ events is 51.1 ± 16.7 which is to be compared with 53.1 predicted by the standard theory for the given running conditions. In terms of the interference cross section discussed above one would expect 108 events with no interference and 163 events with constructive interference. Although the statistical significance of this very difficult experiment is marginal for a serious test of the standard theory (the Weinberg angle, e.g. can only be determined with a relative error of about 50%), the constructive interference of CC and NC is excluded by more than 4 standard deviations. New, still preliminary data from this experiment have been presented at the Munich Conference (Talaga, 1988), supporting the previous findings. In total, 265 ± 52 events have been found after background subtraction, corresponding to a cross section

$$\frac{1}{E}\,\sigma\,(\nu_e\,e \rightarrow \nu_e\,e) = (11.2 \pm 2.2 \pm 1.2) \times 10^{42}\left[\frac{\text{cm}^2}{\text{GeV}}\right],$$

in good agreement with the expectation from the standard theory with $\sin^2\theta_W = 0.23$ (see Fig. 4.3).

4.3 Neutrino-Nucleon Scattering

Interpreting physics results from neutrino interactions on nucleons in terms of elementary neutrino quark scattering is burdoned with a number of theoretical complications, in contrast to the νe reactions discussed in the previous chapter. First of all, free quark targets do not exist. Quarks are believed to be the constituents of the nucleon, where they are held together by the colour forces. These forces are strong at large distances and thus ensure the binding of quarks inside the hadrons. But they become weak at small distances or, equivalently, at large momentum transfers Q^2. This property of the colour force, called "asymptotic freedom" as Q^2 approaches infinity, is evident in the form of the running coupling α_s (see (3.64)).

Asymptotic freedom was anticipated with the simple quark-parton model (QPM) as introduced by Feynman (1969). In this model the quarks inside the nucleon are treated as quasi-free and non-interacting. The scattering process can then be viewed according to the graph in Fig. 4.7, where one of the three quarks is struck by the probing current (here W or Z), while the other two quarks are acting as "spectators". At high momentum transfers to the struck quark, the three-quark system usually has a large invariant mass and decays into a large number of secondary hadrons (deep inelastic scattering).

In the framework of non-abelian gauge theories such as QCD, as noted above, there is a natural explanation for the approximate validity of the QPM in deep inelastic scattering in terms of the running coupling constant, which, under certain circumstances, decreases as the Q^2 increases. However, QCD also introduces definite modifications to the QPM which are related to the QCD gauge fields, i.e. the gluons. Some representative examples of diagrams incorporating gluons are shown in Fig. 4.8.

The fact that the target quarks are trapped inside the nucleons has two basic consequences for the physical analysis of the scattering process:

Target momentum unknown: The target quarks are not at rest. Each quark carries a certain fraction x of the total nucleon momentum. The corresponding parton distribution functions $q_i(x)$, $i = u$, d and their dependence on other kinematic variables, such as the momentum-transfer squared Q^2 from the incoming neutrino to the outgoing lepton, are a priori unknown. Following Bjorken (1976),

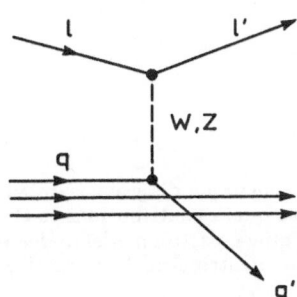

Fig. 4.7. Feynman diagrams for deep-inelastic neutrino-nucleon scattering in the simple quark-parton model

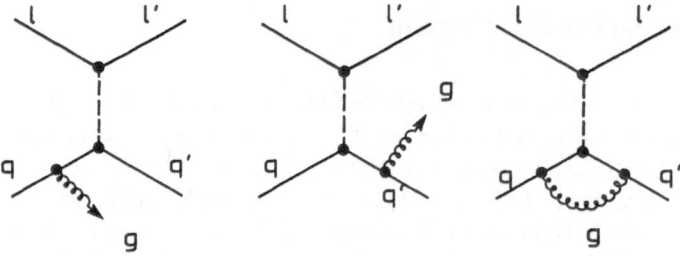

Fig. 4.8. Feynman diagrams symbolising QCD corrections to the simple quark-parton model in deep-inelastic neutrino-nucleon scattering

the scaling variable x is given by $x = Q^2/2M\nu$, where M is the nucleon mass and ν the energy transfer from initial to final lepton. Scaling violations of the parton distribution functions, i.e. their dependence on Q^2, are established experimentally (see e.g. Dydak, 1983 or Sciulli, 1985 for reviews) and explained by QCD corrections (gluon radiation, see Fig. 4.8) to the QPM. Numerous calculations of these QCD corrections exist and one may consult, e.g., the articles by Politzer (1977), Sachrajda (1978), Ellis et al. (1978), or Duke and Roberts (1984) for further details.

Target flavour unknown: For charged current (CC) reactions the flavour of the target quark is known, whereas for neutral current (NC) reactions it is not. Thus in νN interactions mediated by the neutral current, where N stands for proton or neutron, one will always measure the sum of contributions from u and d quarks. The QCD corrections to the QPM mentioned above also allow for a $q\bar{q}$ sea (see Fig. 4.9), so that flavours other than u, d (valence quarks) may contribute to the neutral current cross section. The sea contribution is expected to be sizeable only for small x (≤ 0.2). Furthermore, the scattered quarks are not observable as free particles (although unconfirmed evidence for free quarks has been reported in the past by La Rue et al. (1981)). The scattered quark and the "spectator" diquark system, and possibly additional gluons, convert to hadrons which are finally measured in a detector. As in the case of e^+e^- annihilation into hadrons, parton fragmentation functions $D_i^h(z)$ have to be introduced which may be modelled according to various prescriptions. The functions $D_i^h(z)$ are a measure for the probability for the parton i to emit a given hadron h with momentum fraction z of the parent parton.

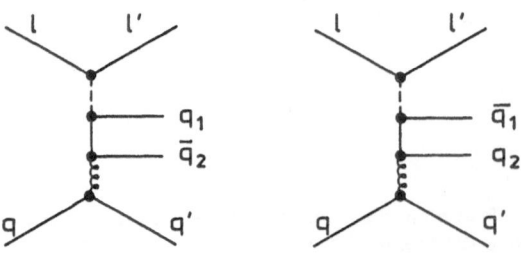

Fig. 4.9. Feynman diagrams symbolising sea quark contributions to the simple quark-parton model in deep-inelastic neutrino-nucleon scattering

Finally, the extraction of quark couplings to the weak neutral current heavily relies on the assumption that the quark distribution functions inside the nucleons do not depend on the probing current. In particular, the quark distribution functions and their Q^2 evolution, derived from charged current reactions (W exchange) and charged lepton scattering (γ exchange), both for valence and sea quarks, are used to establish relations between ν cross sections and the weak neutral couplings of the quarks (see below).

Neutral current interactions of neutrinos and quarks at present accelerator energies can be described in the local limit, using the effective Lagrangian (2.35)

$$\mathcal{L}_{\text{eff}}^{\nu q} = -\frac{G}{\sqrt{2}}\, \bar{\nu}\gamma_\lambda\,(1-\gamma_5)\nu \Big[u_{\text{L}}\, \bar{u}\gamma_\lambda\,(1-\gamma_5)\,u + u_{\text{R}}\, \bar{u}\gamma_\lambda\,(1+\gamma_5)\,u$$

$$+\, d_{\text{L}}\, \bar{d}\gamma_\lambda\,(1-\gamma_5)\,d + d_{\text{R}}\, \bar{u}\gamma_\lambda\,(1+\gamma_5)\,d \;\; + \ldots \; \Big],$$

(4.17)

where the dots represent $s, c, b \ldots$ sea quarks. The chiral couplings $u_{\text{L}}, u_{\text{R}}$ etc. are given by (2.20) and Table 2.1, where, e.g. , c_{L} for the u quark is now denoted by u_{L} for simplicity.

Although expressions for the couplings will be used which are suggested by the standard theory, the Lagrangian (4.17) is general enough to apply to any model with vector and axial-vector currents. These couplings may also represent the effects of several neutral weak bosons (Hung and Sakurai, 1977). For the subsequent development of the relevant formulae the sea quarks will not be considered. Their contribution is small and can be corrected for within the assumption outlined above.

The data on νq neutral current interactions essentially come from deep-inelastic scattering on proton, neutron and isoscalar targets (equal number of u and d quarks), semi-inclusive pion production, and exclusive channels such as elastic scattering and single pion production. Rather than presenting in detail the individual experiments and their results, which has been comprehensively done elsewhere (see, e.g., Sciulli 1979; Kim et al., 1981; Myatt, 1982; Barbiellini and Santoni, 1986), only the essential steps leading to a unique determination of the weak neutral couplings of the u and d quarks will be covered. Since there are, however, new high precision data available not yet covered in the above cited works, some discussion will be devoted to these new experiments and their analyses.

Similar to the e^+e^- experiments and νe data, deep inelastic neutrino quark scattering can be analysed in various ways. As suggested by the effective Lagrangian (4.17) one may investigate the allowed domains for the weak charges v and a in a quasi model-independent fashion and compare to the definite predictions of the standard theory. Alternatively the standard theory expressions for v and a can be used to determine the "only" free parameter of the theory $\sin^2\theta_{\text{W}}$. Both of these procedures will be outlined in the following, starting with the model-independent analyses.

4.3.1 Deep-inelastic Scattering on Isoscalar Targets

In the simple quark model the cross section for NC and CC reactions can be easily obtained with the helicity amplitudes already used for νe scattering. The major difference will be to replace the electron by a quark, the momentum distribution of which being characterised by the parton distribution functions $q_i(x)$, $i = u, d$. The momentum carried by a parton i in a given collision is therefore $x\, q_i(x)$. Clearly, the average parton momentum in the collision process is obtained by integration over x. For an isoscalar target (equal numbers of protons and neutrons inside the target nucleus) both u and d quarks contribute with equal weights so that an average nucleon structure function $F(x)$ can be defined which reads

$$F(x) = \frac{1}{2}\, x\, (q_u(x) + q_d(x)),\tag{4.18}$$

where $q_u(x)$ and $q_d(x)$ are the distribution functions of u and d inside the proton (note that by isospin invariance $q_u(x)$ inside the proton is the same as $q_d(x)$ inside the neutron). With the above structure function $F(x)$ the differential cross section per nucleon for neutrino scattering on an isoscalar target is obtained (in analogy to (4.6,7), see also Sehgal (1977)) as follows:

$$\frac{d\sigma_\nu^{NC}}{dy} = \frac{2\varrho^2\, G^2\, M E_\nu}{\pi} \int F(x)\left[(u_L^2 + d_L^2) + (u_R^2 + d_R^2)(1 - y)^2\right] dx$$

$$\frac{d\sigma_\nu^{CC}}{dy} = \frac{2\, G^2\, M E_\nu}{\pi} \int F(x)\, dx \tag{4.19}$$

and

$$\frac{d\sigma_{\bar\nu}^{NC}}{dy} = \frac{2\varrho^2\, G^2\, M E_{\bar\nu}}{\pi} \int F(x)\left[(u_R^2 + d_R^2) + (u_L^2 + d_L^2)(1 - y)^2\right] dx$$

$$\frac{d\sigma_{\bar\nu}^{CC}}{dy} = \frac{2 G^2 M E_{\bar\nu}}{\pi} \int F(x)\,(1 - y)^2\, dx. \tag{4.20}$$

In the above equations, M is the mass of the target nucleon, and the scaling variable y is the energy loss between incoming and outgoing lepton in the laboratory system, in units of the incoming neutrino energy

$$y = 1 - \frac{E}{E_\nu}.\tag{4.21}$$

One can get rid of the unknown structure functions $F(x)$ by integrating over x and taking ratios between NC and CC cross sections. As in the case of neutrino electron scattering, systematic erros due to flux normalisation cancel in the cross section ratios. The total cross sections are obtained by further integrating over y. For the ratios one then finds

$$R_\nu = \frac{\sigma_\nu^{NC}}{\sigma_\nu^{CC}} = \left[u_L^2 + d_L^2 + \frac{1}{3} \left(u_R^2 + d_R^2 \right) \right] \varrho^2$$

$$R_{\bar\nu} = \frac{\sigma_{\bar\nu}^{NC}}{\sigma_{\bar\nu}^{CC}} = \left[u_L^2 + d_L^2 + 3 \left(u_R^2 + d_R^2 \right) \right] \varrho^2 \,.$$

(4.22)

These relations are rigorously true only for integration over the full x and y ranges. Experimentally, this would correspond to measuring all NC and CC events down to the minimal momentum transfer between incoming and outgoing lepton. Taking the example of NC events, it is evident that for the detection, proper classification and efficient discrimination against background a lower bound for the recoil hadronic energy has to be imposed. A typical cut in the visible hadronic energy is of order 5 GeV. As a consequence of this requirement not only the y interval is affected but also the integration interval for the scaling variable x in (4.19, 20). The limited range in x affects the integrals over $q_u(x)$ and $q_d(x)$ in different ways, since these distribution functions differ substantially from each other. This has been shown, e.g., in deep-inelastic electron nucleon scattering carried out by the SLAC-MIT Collaboration (Bockk et al., 1979) and in deep-inelastic muon scattering by the European Muon Collaboration (Aubert et al., 1983). In an actual experiment the effects due to the limited integration intervals are customarily taken into account by introducing correction coefficients which multiply the individual chiral couplings in (4.22). These coefficients furthermore account for other experimental and theoretical corrections, such as efficiencies due to the event selection criteria, neutrino fluxes, deviations from isoscalarity of the target nucleus, sea quark effects and higher order QCD corrections to the nucleon structure functions.

By linearly combining R_ν and $R_{\bar\nu}$ from (4.22) one can determine $\varrho^2(u_L^2 + d_L^2)$ and $\varrho^2(u_R^2 + d_R^2)$ which leads to allowed regions for the chiral couplings within concentric rings in the (u_L, d_L) and (u_R, d_R) plane, respectively. The effect of the coefficients multiplying the chiral couplings is to slightly modify the shape of the annular regions. Kim et al. (1981) have determined these regions from the data then available, using detailed phenomenological parametrisations of the parton distribution functions to account for gluon radiation and sea quark effects (see Buras and Gaemers, 1978). The experimental data came from the Harward-Pennsylvania-Wisconsin-Fermilab Collaboration (HPWF, Wanderer et al., 1978), the Caltech-Illinois-Tufts-Fermilab Collaboration (CITF, Meritt et al., 1978), the Aachen-Bonn-CERN-London-Oxford-Saclay Collaboration (AB-CLOS, Deden et al., 1979), and the CERN-Heidelberg-Dortmund-Saclay Collaboration (CDHS, Geweniger et al., 1979). Since then further experiments have been carried out to measure both R_ν and $R_{\bar\nu}$ using either bubble chamber techniques, e.g., the Aachen-Bonn-CERN-Democritos-Imperial College-London-Oxford-Saclay Collaboration (Bosetti et al., 1983), the Amsterdam-Bergen-Bologna-Padova-Pisa-Saclay-Torino Collaboration (Allasia et al., 1983) the Stony Brook-Illinois-Maryland-Tokyo-Tufts Collaboration (SIMTT, Kafka et al., 1982), and the SKAT Collaboration at Serpukhov (Ammosov et al., 1986), or counter techniques such as the CHARM Collaboration (Jonker et al., 1981), the CCFRR

Table 4.4. Measurements of the ratios R_ν and $R_{\bar\nu}$ of NC to CC reactions in deep-inelastic neutrino interactions on isoscalar targets

Experiment	E^h_{cut} [GeV]	R_ν	$R_{\bar\nu}$
HPWF (Wanderer et al., 1978)	4.0	0.30 ± 0.04	0.33 ± 0.09
CITF (Merritt et al., 1978)	4.0	0.28 ± 0.03	0.35 ± 0.11
CHARM (Jonker et al., 1981)	2.0	0.320 ± 0.010 ± 0.003	0.377 ± 0.020 ± 0.003
BEBC (Ne/H$_2$) (Bosetti et al., 1983)	9.0	0.345 ± 0.015 ± 0.009	0.364 ± 0.029 ± 0.009
BEBC (D$_2$) (Allasia et al., 1983)	5.0	0.328 ± 0.020 ± 0.019	0.353 ± 0.038 ± 0.020
CDHS (Abramovicz et al., 1985)	10.0	0.301 ± 0.005 ± 0.005	0.363 ± 0.010 ± 0.012
SKAT (Freon) (Ammosov et al., 1986)		0.33 ± 0.02	0.44 ± 0.11

Collaboration (Reutens et al., 1985) and the CDHS Collaboration (Abramovicz et al., 1985). CDHS have accumulated the largest event samples with 62 000 ν and 21 000 $\bar\nu$ events, with running conditions optimised for precision in both R_ν and $R_{\bar\nu}$.

A compilation of the experimental values for R_ν and $R_{\bar\nu}$ is presented in Table 4.4, the statistically more significant determinations are shown in Fig. 4.10. Also drawn in this figure is the expectation for R_ν, $R_{\bar\nu}$ within the framework of the standard theory, as a function of the value for $\sin^2\theta_W$. This functional relation, which allows the precise determination of $\sin^2\theta_W$ from $R_\nu(R_{\bar\nu})$, will be used below. For the moment, the extraction of the chiral couplings $u_{L,R}$ and $d_{L,R}$ from neutrino-nucleon scattering in a model-independent way will be emphasized.

As is evident from (4.22), the measurement of R_ν, $R_{\bar\nu}$ allows a determination of $(u_L^2 + d_L^2)$ and $(u_R^2 + d_R^2)$. The 90% C.L. regions for these two combinations, based on the data listed in Table 4.4, are shown in Figs. 4.11,12. The regions are enclosed between the concentric circles around $(u_L, d_L = 0)$ and $(u_R, d_R = 0)$, respectively. These regions have been obtained by Panman (1984), who has performed an analysis similar to Kim et al. (1981), including the data from Table 4.4 as well as other neutrino data to be discussed subsequently to further constrain the chiral couplings. Since the overall sign of the Lagrangian (4.17) is

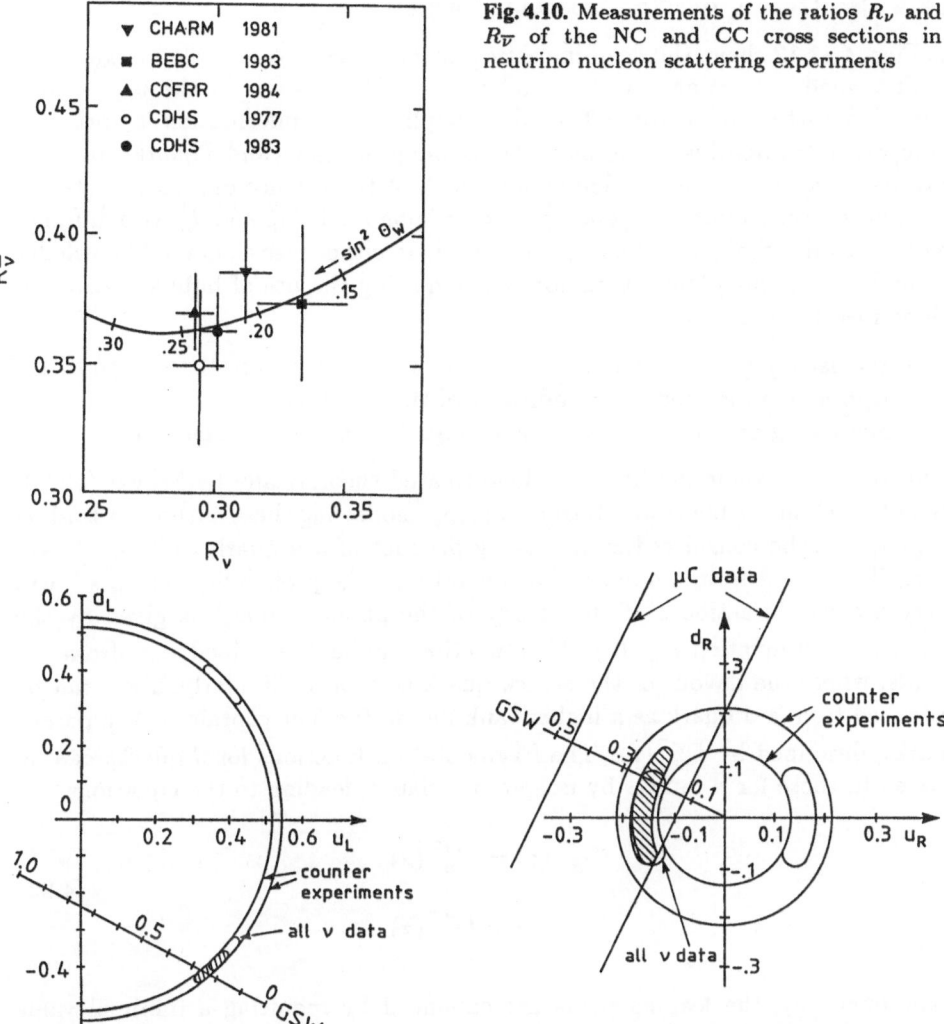

Fig. 4.10. Measurements of the ratios R_ν and $R_{\overline{\nu}}$ of the NC and CC cross sections in neutrino nucleon scattering experiments

Fig. 4.11. 90% C.L. contours for the left-handed couplings of u and d quarks from νN scattering data, using deep-inelastic scattering data from isoscalar targets and from protons, determined by Panman (1984). The shaded area results from a similar analysis by Amaldi et al. (1987)

Fig. 4.12. 90% C.L. contours for the right-handed couplings of u and d quarks from νN scattering data, using deep-inelastic scattering data from isoscalar targets and from protons, determined by Panman (1984). The shaded area results from a similar analysis by Amaldi et al. (1987)

arbitrary, one may choose the sign of a particular chiral coupling , e.g. $u_L > 0$, as is indicated in the figure. This choice is motivated by the standard theory (see Table 2.2), but also independently supported by factorisation relations (see Chap. 7). As noted by Hung and Sakurai (1981), the sign is in principle measurable by interference with gravity, which does not look very promising, however, from an experimental point of view.

4.3.2 Scattering on Non-isoscalar Targets

As Figs. 4.11,12 show, the deep-inelastic scattering data on isoscalar targets give a fairly good determination of the allowed annular regions in the $u_{L,R} - d_{L,R}$ planes, in particular for the left-handed couplings. No information is obtained, however, on the relative strengths of the couplings for the u and d quarks, because isoscalar targets provide no clue about which of the valence quarks was struck. To achieve a separation of u_L^2 and d_L^2 or, correspondingly, u_R^2 and d_R^2, two different experimental techniques have been employed. In either case a detailed knowledge of the hadronic final state is mandatory, requiring the use of bubble chambers. These measurements are:

— Inclusive pion production from isoscalar targets, where the charge of the leading pion is used as an indicator of the quark flavour.
— Scattering on non-isoscalar targets, e.g. on protons and neutrons.

Inclusive pion production has been investigated theoretically by Sehgal (1975), Hung (1977), and Abbott and Barnett (1978). Following these authors, a leading π^+, e.g., can be considered as the decay product of a u quark while a π^- will more likely result from a d quark fragmentation. The probability that a π^+ will carry a certain fraction z of the energy of the parent u quark is given by the fragmentation function $D_u^{\pi^+}(z)$. This function can be determined, e.g., from CC events, where the flavour of the struck quark is known. Similarly, a π^- can be produced from a u quark as a higher rank meson (i.e. not containing the parent quark), described by $D_u^{\pi^-}(z)$. The fragmentation functions for d quarks can be related to those for u quarks by isospin invariance, leading to the equations

$$D_d^{\pi^-}(z) = D_u^{\pi^+}(z)$$

$$D_d^{\pi^+}(z) = D_u^{\pi^-}(z)$$

(4.23)

Experimentally, the leading pions are enhanced by requiring a minimal value for z which excludes softer, higher-rank mesons. Resonances, not accounted for in the QPM, usually lead to few, energetic particles in the final state and can therefore effectively be suppressed by imposing an upper limit for z.

The ratio η of the positive to negative pion yield from u quarks has been determined from both electro-production (Dakin and Feldman, 1975) and CC reactions (Haguenauer, 1974) with the result

$$\eta = \int_{0.3}^{0.7} D_u^{\pi^+}(z)\, dz \;\bigg/\; \int_{0.3}^{0.7} D_u^{\pi^-}(z)\, dz \;=\; 2.8 \pm 0.7 \,.$$

For NC reactions induced by ν's, the ratio of π^+/π^- in the QPM is given by

$$\left(\frac{\pi^+}{\pi^-}\right)_\nu = \frac{(u_L^2 + \frac{1}{3} u_R^2)\, D_u^{\pi^+} + (d_L^2 + \frac{1}{3} d_R^2)\, D_d^{\pi^+}}{(u_L^2 + \frac{1}{3} u_R^2)\, D_u^{\pi^-} + (d_L^2 + \frac{1}{3} d_R^2)\, D_d^{\pi^-}} \,.$$

Table 4.5. Measurements of the ratio π^+/π^- for semi-inclusive pion production in neutrino and anti-neutrino NC interactions. Also given is the expectation from the standard theory, using (4.24) with $\sin^2\theta_W = 0.23$

Experiment	$(\pi^+/\pi^-)_\nu$	$(\pi^+/\pi^-)_{\bar\nu}$
GGM-PS (Kluttig et al., 1977)	0.77 ± 0.14	1.65 ± 0.33
BEBC (Deden et al., 1979a)	0.69 ± 0.22	1.37 ± 0.31
Fermilab 15' (Roe et al., 1979)		1.27 ± 0.22
expectation	0.849	0.985

The corresponding ratio for $\bar\nu$-induced reactions is obtained by interchanging L and R. Observing the isospin relations (4.23) and substituting for η one obtains

$$\left(\frac{\pi^+}{\pi^-}\right)_\nu = \frac{(u_L^2 + \frac{1}{3} u_R^2)\,\eta + (d_L^2 + \frac{1}{3} d_R^2)}{(u_L^2 + \frac{1}{3} u_R^2)\ + (d_L^2 + \frac{1}{3} d_R^2)\,\eta}$$

$$\left(\frac{\pi^+}{\pi^-}\right)_{\bar\nu} = \frac{(u_R^2 + \frac{1}{3} u_L^2)\,\eta + (d_R^2 + \frac{1}{3} d_L^2)}{(u_R^2 + \frac{1}{3} u_L^2)\ + (d_R^2 + \frac{1}{3} d_L^2)\,\eta}\ . \tag{4.24}$$

Since in these relations u_L^2 and d_L^2 (u_R^2 and d_R^2) are multiplied with different coefficients, in contrast to (4.22), the individual squares of the chiral coupling constants can be determined separately. Measurements of pion ratios in neutral current reactions have been published by Kluttig et al. (1977), Deden et al. (1979a), and Roe (1979). They are shown in Table 4.5. For the evaluation of the moduli of the chiral couplings using (4.24), factors of order unity to correct for the neglection of sea quarks, similar to (4.22), need to be introduced. The precision of the data does, however, not warrant a full QCD analysis. The measurements of the pion ratios are in agreement with the naive expectation from the standard theory, which is also given in the table. It should be pointed out, however, that the theoretical uncertainties in deriving (4.24) may be quite large. In particular, the fragmentation functions, assumed to only depend on z, are found to also vary with x and Q^2 (Blietschau et al., 1979).

A technique less sensitive to the hadronisation process is provided by deep-inelastic scattering on protons and neutrons. With the assumption of isospin invariance of the valence quark distribution functions, i.e.

$$x\, q_u(x) = 2\, x\, q_d(x) \equiv 2\, F(x)$$

the following cross sections for ν beams are obtained in the QPM:

$$\sigma^{NC}(\nu p) = \frac{2\varrho^2 G^2 M E_\nu}{\pi} \int dx\, F(x)\left[(2\,u_L^2 + d_L^2) + \frac{1}{3}\,(2\,u_R^2 + d_R^2)\right]$$

$$\sigma^{CC}(\nu p) = \frac{2\,G^2 M E_\nu}{\pi} \int F(x)\, dx$$

$$\sigma^{NC}(\nu n) = \frac{2\varrho^2 G^2 M E_\nu}{\pi} \int dx\, F(x)\left[(u_L^2 + 2d_L^2) + \frac{1}{3}\,(u_R^2 + 2d_R^2)\right] \tag{4.25}$$

$$\sigma^{CC}(\nu n) = \frac{2\,G^2 M E_\nu}{\pi} \int 2\,F(x)\, dx\,,$$

where the factor of 2 in the CC νn cross section with respect to the νp cross section is due to the presence of two d quarks inside the neutron.

The corresponding expressions for antineutrino scattering are obtained by interchanging the indices L and R for the neutral current reactions and attaching a factor of $1/3$ to the CC cross sections from the y integration (see (4.20)):

$$\sigma^{NC}(\bar\nu p) = \frac{2\varrho^2 G^2 M E_{\bar\nu}}{\pi} \int dx\, F(x)\left[\frac{1}{3}(2\,u_L^2 + d_L^2) + (2\,u_R^2 + d_R^2)\right]$$

$$\sigma^{CC}(\bar\nu p) = \frac{2\,G^2 M E_{\bar\nu}}{\pi} \int \frac{1}{3}\,2\,F(x)\, dx$$

$$\sigma^{NC}(\bar\nu n) = \frac{2\varrho^2 G^2 M E_{\bar\nu}}{\pi} \int dx\, F(x)\left[\frac{1}{3}(u_L^2 + 2d_L^2) + (u_R^2 + 2d_R^2)\right] \tag{4.26}$$

$$\sigma^{CC}(\bar\nu n) = \frac{2\,G^2 M E_{\bar\nu}}{\pi} \int \frac{1}{3}\,F(x)\, dx\,.$$

From the relations (4.25, 26) and the values for the chiral couplings in Table 2.2 one expects the ratio R of NC to CC reactions of about $1/2$ for $R(\nu p)$ and $R(\bar\nu n)$ and of about $1/4$ for $R(\bar\nu p)$ and $R(\nu n)$. The data on these processes come exclusively from experiments using either the Big European Bubble Chamber (BEBC) at CERN or the $15'$ chamber at Fermilab. Most experiments were carried out using hydrogen fillings exposed to wide band ν_μ or $\bar\nu_\mu$ beams. Similar to the counter experiments, cuts on the visible hadronic total and transverse momentum are imposed in the event selection in order to discriminate against background. Charged-current events are identified with the help of an external muon identifier (EMI), a counter system surrounding part of the downstream outer walls of the bubble chamber. The available measurements for $R(\nu p)$ are given in Table 4.6.

Traditionally, data for νn interactions were extracted from neon-hydrogen or pure neon bubble chamber experiments (BEBC, Deden et al., 1979 ; FNAL $15'$, Marriner et al., 1977), separating proton and neutron reactions according to the

Table 4.6. Measurements of the ratios R_ν and $R_{\bar\nu}$ for NC to CC reactions in neutrino interactions on protons

Experiment	E^h_{cut} [GeV]	R_ν	$R_{\bar\nu}$
FNAL 15' (H$_2$) (Harris et al., 1977)	10.0	0.48 ± 0.17	
CERN BEBC (H$_2$) (Blietschau et al., 1979a)	$p^h_t > 1.5$	0.51 ± 0.04	
FNAL 15' (H$_2$) (Carmony et al., 1982)	5.0		0.36 ± 0.06
FNAL 15' (D$_2$) (Kafka et al., 1982)	10.0	0.49 ± 0.06	
CERN BEBC (D$_2$) (Allasia et al., 1983)	5.0	0.49 ± 0.05	0.26 ± 0.04
CERN BEBC-TST (H$_2$) (Armenise et al., 1983)	–	0.47 ± 0.04	
CERN BEBC-TST (H$_2$) (Moreels et al., 1984)	–		0.33 ± 0.04
CERN SPS BEBC (H$_2$) (Jones et al., 1986)	–	0.384 ± 0.028	0.338 ± 0.021

Table 4.7. Measurements of the ratios R_ν and $R_{\bar\nu}$ for NC to CC reactions in neutrino interactions on neutrons

Experiment	E^h_{cut} [GeV]	R_ν	$R_{\bar\nu}$
FNAL 15' (D$_2$) (Kafka et al., 1982)	10.0	0.22 ± 0.03	
CERN BEBC (D$_2$) (Allasia et al., 1983)	$(p^h_t > 1.5)$	0.25 ± 0.02	0.57 ± 0.09

charge of the leading pion. This method obviously suffers from the same theoretical uncertainties discussed for the inclusive pion data. When high energy, high intensity (wide band) neutrino beams from the SPS (or at Fermilab) became available, bubble chamber experiments were feasible again using the classical light liquids hydrogen and, in particular for the study of neutrino-neutron interactions, deuterium. As an example, the ABBPPST Collaboration (Allasia et al., 1983) have exposed BEBC, filled with deuterium, to neutrino and antineutrino wide band beams supplied by the CERN SPS. The identification of neutrino neutron reactions proceeds along the following criteria: Reactions on neutrons will lead to final states with no net charge and can thus be recognised by counting the number of energetic particles in the chamber. If this number is even, excluding a possible low momentum spectator proton, the reaction is classified as a neutron target event. Conversely, an event with an odd number of energetic charged particles signals a neutrino proton interaction. Both charged- and neutral-current data are recorded in the same experiment. Systematic errors originating from the event selection procedure are largely reduced by defining the ratio

$$ r = \frac{\sigma^{\mathrm{NC}}(\nu p) - \sigma^{\mathrm{NC}}(\nu n)}{\sigma^{\mathrm{CC}}(\nu p) - \sigma^{\mathrm{CC}}(\nu n)} , $$

which is given by the QPM as

$$ r^{\nu} = (-u_{\mathrm{L}}^2 + d_{\mathrm{L}}^2 - \frac{1}{3} u_{\mathrm{R}}^2 + \frac{1}{3} d_{\mathrm{R}}^2)\, \varrho^2 $$

$$ r^{\bar{\nu}} = (\ \ u_{\mathrm{L}}^2 - d_{\mathrm{L}}^2 + 3 u_{\mathrm{R}}^2 - 3 d_{\mathrm{R}}^2)\, \varrho^2 . $$

(4.27)

Similar to the NC/CC cross section ratios from isoscalar targets R (see (4.22)), the relations (4.27) receive correction coefficients to account for the specific experimental conditions and for the neglect of the sea quarks. Using these data together with the constraints from deep-inelastic neutrino scattering on isoscalar targets presented in the previous section allows a determination of the squares of the individual chiral couplings.

In a recent analysis, updating the pioneering work of Kim et al. (1981), Fogli (1985) has obtained the following values:

$$ u_{\mathrm{L}}^2 = 0.141 \pm 0.019 $$

$$ d_{\mathrm{L}}^2 = 0.170 \pm 0.020 $$

$$ u_{\mathrm{R}}^2 = 0.017 \pm 0.009 $$

$$ d_{\mathrm{R}}^2 = 0.008 \pm 0.008 . $$

(4.28)

The quoted errors reflect both the statistical and systematic uncertainties of the individual experiments, added in quadrature. As a novelty, electroweak radiative corrections, following the approach of Marciano and Sirlin (1980), have been applied in this analysis. Both solutions for the squares of the chiral couplings, with and without radiative corrections, agree within one standard deviation. Similar values for the squares of the chiral couplings have been obtained by Panman (1984), whose results are shown in Figs. 4.11,12. The 90% C.L. intervals for the chiral couplings are now, after inclusion of the scattering data on non-isoscalar targets, confined to two regions mirrored around $u_L = 0$ and $d_R = 0$ for the left- and right-handed couplings, respectively. One of these regions, with $d_L < 0$, is favoured by the standard theory. Further constraints to resolve the remaining twofold ambiguity in the (u_L, d_L) and (u_R, d_R) planes will be provided by exclusive neutrino-nucleon reactions.

4.3.3 Exclusive Reactions

In order to single out one of the two solutions in the left-handed and right-handed planes, physical observables have to be found which depend on linear relations between the couplings, and therefore allow to determine their relative signs. For that purpose it is instructive to re-arrange terms in the effective Lagrangian (4.17), as proposed by Hung and Sakurai (1977), where the weak neutral hadronic current is decomposed into a sum of vector and axial-vector parts with fixed isospin I, $(I = 0, I = 1)$:

$$
\begin{aligned}
\mathcal{L}_{\text{eff}}^{\nu q} &= -\frac{\varrho\, G}{\sqrt{2}}\, \bar{\nu}\gamma^\lambda(1 - \gamma_5)\, \nu \Big\{ \frac{1}{2}\Big[\alpha\left(\bar{u}\gamma_\lambda u - \bar{d}\gamma_\lambda d\right) + \beta\left(\bar{u}\gamma_\lambda\gamma_5 u - \bar{d}\gamma_\lambda\gamma_5 d\right)\Big] \\
&\quad + \frac{1}{2}\Big[\gamma\left(\bar{u}\,\gamma_\lambda\, u + \bar{d}\,\gamma_\lambda\, d\right) + \delta\left(\bar{u}\,\gamma_\lambda\,\gamma_5\, u + \bar{d}\,\gamma_\lambda\,\gamma_5\, d\right)\Big]\Big\} \\
&= -\frac{\varrho\, G}{\sqrt{2}}\, \bar{\nu}\,\gamma^\lambda\,(1 - \gamma_5)\, \nu\Big[\alpha V_\lambda^{I=1} + \beta A_\lambda^{I=1} + \gamma V_\lambda^{I=0} + \delta A_\lambda^{I=0}\Big]\,.
\end{aligned}
\tag{4.29}
$$

The current couplings $\alpha, \beta, \gamma, \delta$ in front of the vector (V) and axial-vector (A) pieces are related to the chiral couplings in the following way:

$$
\begin{aligned}
V, I = 1 &: \quad \alpha = u_L - d_L + u_R - d_R \\
A, I = 1 &: \quad \beta = -u_L + d_L + u_R - d_R \\
V, I = 0 &: \quad \gamma = u_L + d_L + u_R + d_R \\
A, I = 0 &: \quad \delta = -u_L - d_L + u_R + d_R\,.
\end{aligned}
\tag{4.30}
$$

The physical significance of the Hung-Sakurai parametrisation is evident: The four coefficients α, β, γ and δ represent the coupling strengths of the weak neutral current to the vector and axial-vector hadronic current with isospins $I = 0$ and $I = 1$, respectively. Since the right-handed couplings are small compared to the left-handed ones (see (4.28)), the current couplings essentially determine the relative sign of d_L with respect to u_L. If, e.g., this relative sign is negative, the absolute values for α and β are large and one would expect a large isovector contribution to the hadronic neutral current, leading to enhanced $\Delta I = 1$ transitions between the initial $(I = 1/2)$ and final hadronic states. A large isoscalar contribution, on the other hand, will lead to the conclusion that u_L and d_L have

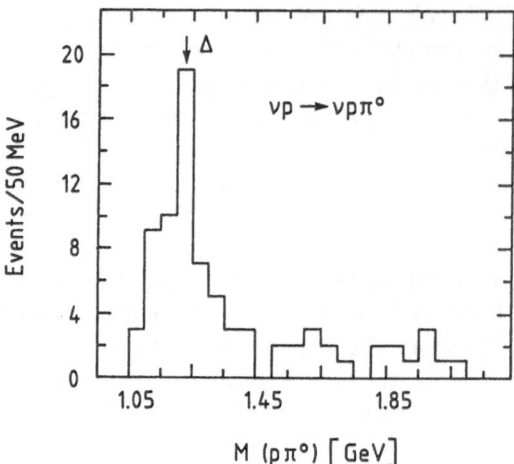

Fig. 4.13. Distribution of the $p\pi^0$ invariant mass in the reaction $\nu_\mu p \rightarrow \nu_\mu p\pi^0$, measured in the Gargamelle heavy liquid bubble chamber (Krenz et al., 1978)

the same sign. In the standard theory, which predicts an odd relative sign, strong isovector contributions in the hadronic transition matrix element are expected.

Single Meson Production: The $\Delta I = 1$ dominance in the hadronic weak neutral current can, e.g., be tested in the reaction $\nu N \rightarrow \nu N \pi$, where strong production of the $I = 3/2$ baryonic resonances, such as $\Delta(1236)$, is expected. Measurements of this reaction in the heavy liquid bubble chamber Gargamelle by Krenz et al. (1978) and Erriquez et al. (1980) have revealed a strong $\Delta(1236)$ signal in the $N\pi$ invariant mass spectrum. As an example, the invariant mass for the combination $p\pi^0$ is shown in Fig. 4.13, where the relative size of the $\Delta(1236)$ signal demonstrates the isovector dominance of the weak hadronic current. These measurements imply $\alpha, \beta \gg \gamma, \delta$. The signs of u_L and d_L must therefore be opposite:

$$u_L\, d_L < 0\,,$$

excluding the positive solution for d_L, which is also incompatible with GSW (see Fig. 4.11).

Another way of exploiting relations (4.30) is to consider the Lorentz structure of the weak neutral current. In analogy to the picture of vector meson dominance in low Q^2 photon-hadron interactions, where the photon couples to quark-antiquark pairs with spin-parity $J^P = 1^-$, the Z^0 may couple to such pairs as well, which then interact strongly with the target nucleus. In the case of Z^0 (or $W^\pm$) exchange, however, both vector and axial-vector parts contribute. The vector part couples predominantly to the ρ meson, while the axial-vector part couples to axial-vector mesons and, by virtue of the PCAC theorem, to pions. These mesons may be detected in the final state by performing coherent-scattering experiments, where the target nucleus interacts as a whole without breakup and does not change the nature of the incoming particle.

Several experiments (Bergsma et al., 1985; Grabosch et al., 1986; Baltay et al., 1986) have recently measured coherent production of π^0 mesons in neutrino (antineutrino) nucleus NC interactions:

$$\nu_\mu(\bar\nu_\mu) + A \to \nu_\mu(\bar\nu_\mu) + A + \pi^0 \,. \tag{4.31}$$

Earlier measurements at the CERN PS have been performed by Faissner et al. (1983) and by the Gargamelle Collaboration (Isiksal et al., 1984). Reaction (4.31) directly probes the isovector axial-vector part of the neutral current, i.e. the current coupling β in (4.30), due to the intrinsic quantum numbers of the π^0. Within the framework of the standard theory one expects (see Table 2.1):

$$\beta = -\left[(u_L - u_R) - (d_L - d_R) \right]$$

$$= -\frac{1}{2}\left[a_u - a_d \right] = -1 \,.$$

A measurement of $|\beta|$ allows a determination of the relative sign of a_u and a_d and consequently a determiation of the relative sign of u_L and d_L, which are large compared to u_R and d_R.

In the coherent production of π^0's the nucleus itself does not break up and therefore only a small amount of recoil energy is transferred to it. The produced π^0's are therefore emitted preferentially close to the forward direction, where they produce narrow electromagnetic showers very similar to those generated by electrons (see also Chap. 4.1). Due to the kinematical similarity with neutrino electron scattering, also here the variable $E\,\theta^2$ is usually studied, where E denotes the energy of the electromagnetic shower generated by the π^0, and θ is the emission angle of the shower relative to the incoming neutrino. A plot of this variable, as measured by the CHARM Collaboration (Bergsma et al., 1985), is shown in Fig. 4.14. The competing backgrounds are neutrino-electron scattering, quasielastic charged current events induced by ν_e ($\bar\nu_e$), incoherent π^0 production, and hadronic events with a large electromagnetic component. These backgrounds are also indicated in the figure. The shapes of the various background contributions have been determined by Monte-Carlo calculations and their relative normalisations were derived from regions where the respective component was expected to dominate. From the observed number of events belonging to coherent π^0 production, cross sections were determined as a function of the incoming neutrino energy. Figure 4.15 shows the cross sections for CHARM and the other quoted experiments together with the theoretical expectation, which is a function of β and was calculated by Rein and Sehgal (1983). The experimental value of $|\beta|$ obtained by CHARM is

$$|\beta| = 1.08 \pm 0.24 \,.$$

In a more direct way, β was determined by the SKAT Collaboration (Grabosch et al., 1986) who measured coherent pion production both in CC ($\pi^\pm$) and NC (π^0) reactions. The relative yield is given by

$$\left(\frac{\pi^0}{\pi^\pm} \right) = \frac{\beta^2}{2} \,.$$

133

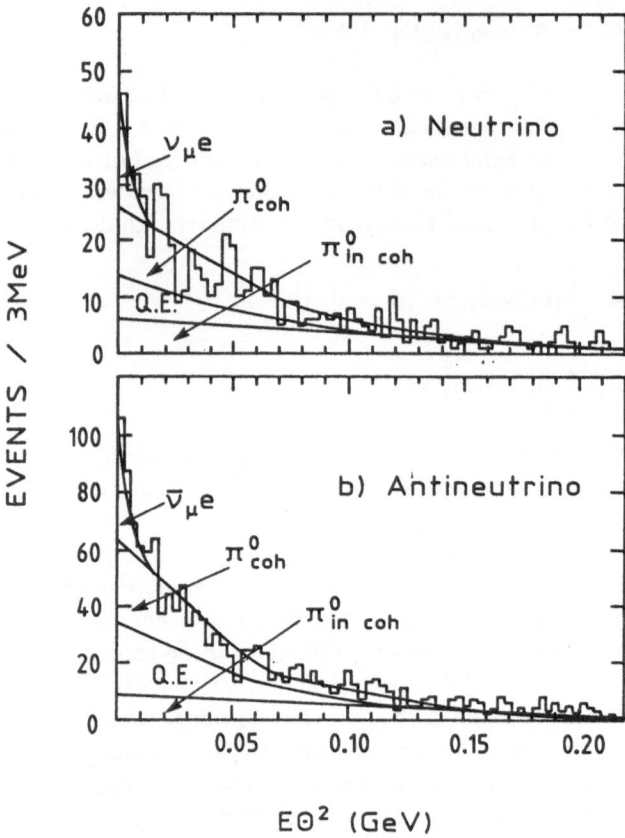

Fig. 4.14. Distributions of the variable $E\theta^2$ for π^0 candidate events from the CHARM experiment (Bergsma et al., 1985). The various contributions are indicated

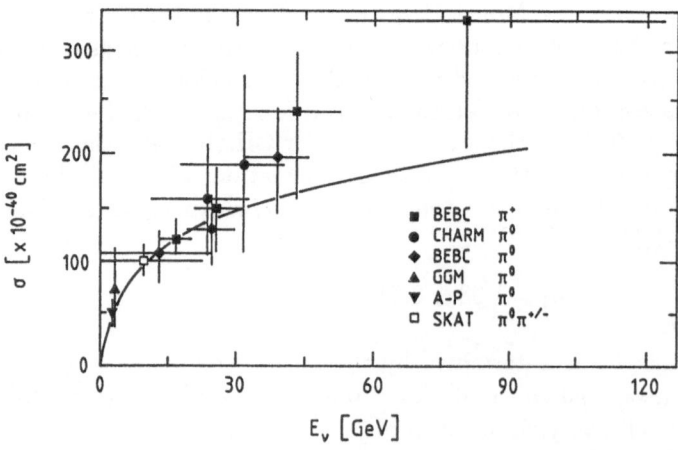

Fig. 4.15. Cross section measurements for coherent pion production in neutrino and antineutrino NC interactions. For comparison, the corresponding cross sections for coherent pion production in CC reactions are also shown. The curve corresponds to the PCAC prediction

Table 4.8. Measurements of $|\beta|$ with coherent π^0 production in neutrino nucleus interactions. The standard theory expectation is $|\beta| = 1$.

Experiment	$|\beta|$
AC-PD (Faissner et al., 1983)	0.93 ± 0.16
CHARM (Bergsma al., 1985)	1.08 ± 0.24
SKAT (Grabosch et al., 1986)	0.99 ± 0.20
Fermilab 15' (Baltay et al., 1986)	0.98 ± 0.24

A compilation of the determinations of the modulus for the axial-vector iso-vector coupling constant β is given in Table 4.8. All values agree nicely with the expectation of $|\beta| = 1$ and thus select from the two solutions for d_L in Fig. 4.11 the one ($d_L < 0$) preferred by the standard theory.

Elastic Scattering: The remaining twofold ambiguity in the right-handed chiral couplings u_R, d_R can be resolved by considering elastic neutrino nucleon scattering. The matrix element for elastic neutrino-proton scattering, written in terms of the vector and axial-vector form factors of the neutral current between proton states (Abbott and Barnett, 1978), is given by

$$\langle p' | J^{\mathrm{NC}}_\lambda | p \rangle = \bar{u}(p') \left[\gamma_\lambda F_1 + \frac{i\sigma_{\lambda\mu}(p-p')^\mu}{2M} F_2 + \gamma_\lambda \gamma_5 F_A \right] u(p) .$$

The vector form factors F_1 and F_2 and the axial-vector form factor F_A can be decomposed into isospin components (Hung, 1978; Abbott and Barnett, 1978)

$$F_i = \alpha F_i^{I=1} + 3\gamma F_i^{I=0}, \quad i = 1, 2$$

$$F_A = \beta F_A^{I=1} + \frac{3}{5}\delta F_A^{I=0} .$$

(4.32)

In these expressions the coefficients α, β, γ, and δ are the same as those given in (4.30), while the form factors $F_i^{I=1}$ and $F_i^{I=0}$ are related to the electromagnetic form factors of the proton and neutron by the conserved vector current (CVC) hypothesis. For the axial-vector form factor one has to resort to specific model assumptions. In the dipole approximation, the Q^2 dependence of the $I = 1$ part can be parametrised as

$$F_A^{I=1} = \frac{1}{2} \frac{1.26}{(1 + Q^2/M_A^2)^2} ,$$

(4.33)

where the dipole mass M_A is close to the mass of the A_1 meson. The isoscalar term cannot contribute, since $\delta = 0$ in the standard theory. With these settings of the form factors involved in (4.32), the coefficients α, β and γ can be determined. Comparing the measured form factors for elastic scattering on protons (Entenberg et al., 1978; Abe et al., 1986) with (4.32) leads to the qualitative result

$$u_L \, u_R < 0 \, ,$$

which favours the domain in the $u_R - d_R$ plane compatible with the standard theory, see Fig. 4.12.

A number of other exclusive ν reactions can be thought of providing additional information about the isospin structure of the weak neutral hadronic current and, by means of (4.30), about the relative signs of the chiral couplings. As examples, the reaction $\bar{\nu}_e \, D \rightarrow \bar{\nu}_e \, p \, n$ (Pasierb et al., 1979) and $\bar{\nu} N \rightarrow \bar{\nu} N \rho^0$ should be mentioned (see Ammosov et al. (1986a) for coherent $\rho^\pm$ production in CC reactions). In most of these reactions, since they have to be carried out in bubble chambers, statistics will be the limiting factor.

The overall agreement of the results from all neutrino nucleon scattering experiments with the relations between the chiral couplings of the quarks of the first generation, as given by the standard theory, is impressive. Amaldi et al. (1987) have recently combined the available data on ν nucleon interactions and have performed a very detailed analysis on the chiral couplings of u and d quarks. They obtain

$$
\begin{aligned}
u_L &= \quad 0.339 \pm 0.017 & d_L &= -0.429 \pm 0.014 \, , \\
u_R &= -0.172 \pm 0.014 & d_R &= -0.011 \pm 0.069 \, .
\end{aligned}
\tag{4.34}
$$

These values are in remarkable agreement with the predictions of the standard theory for $\sin^2 \theta_W = 0.23$, as given in Table 2.2. They are also in excellent agreement with previous analyses such as the one by Fogli (1985), or the one by Kim et al. (1981), who obtained nearly the same central values but substantially larger errors due to the less precise data then available. The allowed regions for the chiral couplings, corresponding to the 95 % C.L. as obtained in the analysis by Amaldi et al., are shown in Figs. 4.11,12 as shaded areas. They are evidently the ones allowed by the standard theory.

4.3.4 Determination of the Weinberg Angle

Theoretical Considerations

As was shown in the previous section, the model-independent analyses of the chiral couplings strongly support the relations among the quark couplings predicted by the standard theory (see Table 2.1). In particular, all ν data seem to be consistent with the value $\sin^2 \theta_W \sim 0.23$ which is, to lowest order, the only additional free parameter in the theory. The next step therefore is to accept

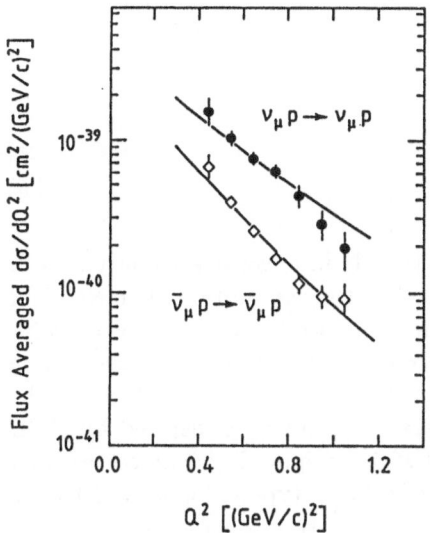

Fig. 4.16. Differential cross section for elastic neutrino nucleon scattering, measured by the E734 Experiment (Abe et al., 1986). The curves through the data are fits using a model for the axial vector form factor

these relations and to determine the Weinberg angle from the data. As for the other reactions discussed in previous chapters, radiative corrections will have to be taken into account. Since the theory is renormalisable these corrections can be calculated, in principle to all orders. Precise determinations of $\sin^2 \theta_W$ from different reactions, such as semileptonic neutrino scattering or direct measurements of the masses of the intermediate vector bosons W and Z (see Chap. 6), will eventually provide important tests of the standard theory at the quantum loop level, comparable to the calculation and measurement of $g - 2$ in QED.

In neutrino scattering experiments, the value of $\sin^2 \theta_W$ is most accurately determined from the ratio R of semileptonic NC to CC cross sections. This was already evident from Fig. 4.10, where the ratio of NC to CC reactions was shown for neutrino and antineutrino induced reactions on isoscalar tagets. Expressing the chiral coupling constants of the u and d quarks in terms of the Weinberg angle and setting $\varrho = 1$ as demanded in the minimal model, the relations (4.22) can be re-written as follows (see Table 2.1):

$$R_\nu = \left(\frac{1}{2} - \sin^2 \theta_W + \frac{20}{27} \sin^2 \theta_W \right)$$

$$R_{\bar{\nu}} = \left(\frac{1}{2} - \sin^2 \theta_W + \frac{20}{9} \sin^2 \theta_W \right). \tag{4.35}$$

As can be seen from Fig. 4.10, the ratio R_ν is more sensitive to $\sin^2 \theta_W$ than $R_{\bar{\nu}}$. It therefore seems advisable to collect data in neutrino rather than antineutrino beams. From a theoretical point of view, relations (4.35) suffer from the neglect of sea quark and gluon contributions as was discussed in Sect. 4.3.1. Llewellyn-Smith (1983) has given improved expressions for $R_{\nu(\bar{\nu})}$ which absorb part of the necessary sea quark corrections to be applied to relations (4.35):

$$\frac{d^2\sigma_{\nu(\bar{\nu})}^{\mathrm{NC}}}{dx\,dy} = \left(\frac{1}{2} - \sin^2\theta_{\mathrm{W}} + \frac{5}{9}\sin^2\theta_{\mathrm{W}}\right)\frac{d^2\sigma_{\nu(\bar{\nu})}^{\mathrm{CC}}}{dx\,dy} + \frac{5}{9}\sin^2\theta_{\mathrm{W}}\frac{d^2\sigma_{\bar{\nu}(\nu)}^{\mathrm{CC}}}{dx\,dy}$$

$$+ \left(\frac{1}{18}\varepsilon\sin^2\theta_{\mathrm{W}} + \left(\frac{1}{4}\sin^2\theta_{\mathrm{W}} - \sin^2\theta_{\mathrm{W}}\right)\tilde{\varepsilon}\right)$$

$$\times \left(\frac{d^2\sigma_{\nu}^{\mathrm{CC}}}{dx\,dy} + \frac{d^2\sigma_{\bar{\nu}}^{\mathrm{CC}}}{dx\,y}\right). \tag{4.36}$$

These relations, which are valid for isoscalar targets in a world with only u, d, $\bar{u}$ and $\bar{d}$ quarks and zero Cabibbo angle, are based on the assumption of an $SU(2)\times U(1)$ structure of the weak neutral current. Making use essentially of isospin invariance, (4.36) expresses the neutral current cross section in terms of the charge current cross section for both neutrino and antineutrino nucleon scattering. The last term in (4.36), which contains the constants ε and $\tilde{\varepsilon}$, is inspired by QCD corrections to the simple quark parton model with $\tilde{\varepsilon}$ originating from constituent quark mass effects ($\tilde{\varepsilon} \sim m_q^2/Q^2$) and ε behaving like a typical higher twist term ($\varepsilon \sim \langle p_{\mathrm{T}}^2\rangle/Q^2$) for large Q^2. Upon integration over the variables x and y and dividing by σ_ν^{CC} one obtains for R_ν:

$$R_\nu = \left(\frac{1}{2} - \sin^2\theta_{\mathrm{W}}\right) + \frac{5}{9}\sin^2\theta_{\mathrm{W}}\left(1 + r\right), \tag{4.37}$$

with

$$r = \sigma_{\bar{\nu}}^{\mathrm{CC}}/\sigma_\nu^{\mathrm{CC}}.$$

The very small corrections expected from the constituent masses of the valence quarks and from higher twist have been omitted in the formula. The theoretical uncertainty of relation (4.37) is reduced with respect to (4.35) owing to the fact that, first of all, only the heavy quark (s, c) contribution to the sea, rather than the contribution from all quarks, have to be corrected for. Secondly, and more importantly, the higher twist terms can be shown, under reasonable assumptions (Llewellyn-Smith, 1983), to result in an uncertainty for the Weinberg angle of $\delta\sin^2\theta_{\mathrm{W}} \leq 0.001$. Neglecting these higher twist terms leads to a total theoretical uncertainty of $\delta\sin^2\theta_{\mathrm{W}} \leq 0.002$, dominated by the s, c quark sea contribution and the incomplete knowledge of the CKM matrix elements.

Thus the theoretical error for the Weinberg angle, as derived from deep-inelastic neutrino nucleon scattering with its complicated and somewhat indirect connection to the basic neutrino quark scattering process, has been significantly improved with the help of relation (4.37), when applied to isoscalar targets (for a detailed discussion see Llewellyn-Smith (1983)). To exploit this strong point requires the measurement of both neutrino and antineutrino CC cross sections in addition to the neutrino NC cross section. However, as the ratio r only enters as a small correction to R_ν (see (4.37), from where one deduces $\delta\sin^2\theta_{\mathrm{W}} \sim \frac{1}{2}\delta r/r$), there is no need for high statistical precision in the antineutrino running.

Equation (4.36) can also be used to deduce another important relation between the Weinberg angle and the cross sections for deep-inelastic neutrino scattering, resulting in a second method of measuring $\sin^2\theta_W$, known as the Paschos-Wolfenstein relation (Paschos and Wolfenstein, 1973):

$$\frac{\sigma_\nu^{NC} - \sigma_{\bar\nu}^{NC}}{\sigma_\nu^{CC} - \sigma_{\bar\nu}^{CC}} = \frac{1}{2} - \sin^2\theta_W . \tag{4.38}$$

Since this relation is formed from differences of neutrino and antineutrino cross sections one immediately recognises its superior theoretical value: In the difference of cross sections the sea quark contributions cancel as well as the contributions from constituent mass and higher twist effects, the remaining uncertainty coming mostly from the off-diagonal elements of the CKM matrix. As discussed in detail by Paschos and Wirbel (1982) the theoretical uncertainty of $\sin^2\theta_W$, when extracted from the data with the Paschos-Wolfenstein relation, is reduced to $\delta\sin^2\theta_W \leq 0.001$. However, in using the Paschos-Wolfenstein relation one is confronted with the experimental problem of limited statistics in antineutrino scattering due to the low flux ratio of $\bar\nu/\nu$, which is typically $1/5$ in high energy narrow band beams. Still, both methods have been successfully used in high statistics neutrino-nucleon scattering experiments to determine the value of $\sin^2\theta_W$.

There is one further theoretical complication related to the CC reactions, common to either relation presented above: Since the charged current does not conserve flavour, charmed quarks may be created in collisions of the neutrino with a valence quark. The relations given in (4.37, 38) tacitly assume that charmed quark production proceeds at full strength, i.e. not suppressed by threshold effects due to the mass of the charmed quark. This assumption is certainly justified at momentum transfers large with respect to the constituent mass. For the present "low energy" experiments with typical $\langle Q^2 \rangle \sim 20$ GeV$^2/c^2$, the zero mass approximation does not hold and charmed quark mass corrections have to be applied according to some model. Charm threshold effects are usually accounted for using the slow rescaling model (Barnett, 1976; Georgi and Politzer, 1976) which raises the effective threshold for charm production substantially above the kinematic threshold. The corresponding correction to $\sin^2\theta_W$ depends on experimental details such as the cut E_{cut}^h in the hadronic energy of the CC events, necessary to discriminate against background. A high cut will select events closer to the asymptotic limit of full charm excitation and therefore lead to a small correction. For the mass of the charmed quark a value around 1.5 ± 0.3 GeV$/c^2$ is usually assumed, the quoted error giving rise to an uncertainty in $\sin^2\theta_W$ of $\delta\sin^2\theta_W \sim 0.004$.

Precise Experiments

Two experiments at CERN have finished their analyses to measure R_ν with high precision: the CDHS Collaboration (Abramowicz et al., 1986) and the CHARM Collaboration (Allaby et al., 1986). Both groups used the same narrow-band muon neutrino beam with 160 GeV central energy. But the experimental set-ups and methods to extract R_ν from the data are sufficiently different so that a comparison of the results will provide a good test of the systematic uncertainties involved. The main difference in the two detector systems is the kind of absorber material used: While the CHARM detector is built from "light" absorber material (marble), the CDHS Collaboration have decided to use iron. Both detectors have briefly been sketched in Sect. 4.1.

Owing to their very different experimental set-ups, the two experiments have developed quite dissimilar event selection philosophies. While it is fairly straight-forward for each experiment to discriminate neutrino interactions from background such as cosmic ray or neutron-induced events, considerable ingenuity is required to distinguish between NC and CC interactions on an event to event basis with known and controllable efficiencies. CDHS have chosen the event length L, defined as the longitudinal distance between the vertex and the last scintillator hit in the event, and the (hadronic) shower energy, derived from the pulse height recorded in the first 1.5 m of iron after the vertex. For each event a cutoff length L_{cut} is calculated which varies longitudinally with the shower energy. Events with $L < L_{cut}$ are attributed to the NC sample, the remaining events are considered as CC reactions. The functional dependence of L_{cut} on the shower energy was determined in such a way as to minimise the spill-over of NC events into the sample of CC events. A minimum of 10 GeV energy deposition was required. For CC events the shower energy also contains the energy deposited by the muon which induces a systematic error in R_ν of $\pm 0.3\%$. Several corrections were applied to the raw data, of which the most important originates from so-called "short" CC events within the NC sample. These are mostly events with large values of y (see (4.21)for the definition), with a correspondingly low momentum muon. Another significant correction comes from events induced by electron neutrinos from K_{e3} decays. These events are classified as NC since the final state electron is hidden in the hadronic shower. The absolute numbers of ν_e induced CC events is determined by Monte Carlo calculations, normalised to the number of ν_μ induced CC events from $K_{\mu 2}$ decays. The ν_e CC events are subtracted from the NC sample and the subclass with a shower energy larger than 10 GeV is added to the CC sample. Thus the CDHS detector gives the ratio R_ν for both ν_μ and ν_e induced reactions. A summary of the most important corrections to the raw data sample for CDHS, and CHARM, to be discussed subsequently, are listed in Table 4.9. The event numbers given are obtained from a total of 5×10^{18} protons of 450 GeV/c momentum delivered on target by the CERN SPS. Not shown in the table are the CC event numbers from about 10^{18} protons on target with the $\overline{\nu}_\mu$ beam.

Due to its fine-grain structure the CHARM neutrino detector allows identifying the muons generated in CC ν_μ nucleon interactions. A muon is defined as a particle penetrating at least 20 calorimeter planes, starting from the event vertex, which corresponds to an energy deposition of at least 1 GeV. The muon is required to originate from the event vertex and has to be visible, i.e. isolated from other tracks, over a range corresponding to an energy loss of at least 2/3 of a GeV. The events satisfying the muon criteria are classified as CC, all others as NC events. A special feature of the CHARM detector is the low detection threshold which makes it possible to trigger on showers down to an energy of

Table 4.9. Event numbers and the most important corrections for NC and CC events in the CDHS and CHARM analyses. $E^h_{cut} = 10$ GeV and 4 GeV for CDHS and CHARM, respectively

	CDHS		CHARM	
	NC	CC	NC	CC
Candidates	60 936	137 853	39 239	108 472
Cosmic rays	−1120	∼ 0	− 557	∼ 0
WBB Backgr.	−2920	−5187	−1753	−4311
Long Shower	+ 159	− 158	+1892	−1835
Short CC	−9642	−9526	−3737	+3735
K_{e3} Corr. (CC)		+2488	−1768	− 106
K_{e3} Corr. (NC)	−3016		− 532	− 33
Corrected	44 397	144 513	32 831	105 982

2 GeV with very high efficiency. Thus in the selection of candidates a threshold for the shower energy of as low as 2 GeV could be imposed. The energy deposition of the muon in the hadron shower region was subtracted from the hadron energy. Besides the standard corrections for cosmic ray and wide band beam (WBB) backgrounds the most important correction comes from the "short" CC events where the muon escaped identification. Some of these events contain muons with an energy less than 1 GeV, and others are caused by muons of more than 1 GeV but leaving the sides of the detector. This very important correction has been determined and cross-checked by independent methods using the data in combination with Monte Carlo techniques. One such method is to overlay a Monte-Carlo-generated muon onto a real NC event, which thus becomes "real" CC, and to determine the fraction of events retained as CC after applying the muon criteria. A complementary correction arises from "long" NC events which contain a track fulfilling the muon criteria and are consequently classified as CC events. These tracks may either originate from decaying pions or kaons in the hadronic shower or from non-interacting (punch-through) hadrons. By analysing CC events in which the identified muon has been removed, this correction can be estimated from the number of events where a second track satisfies the muon criteria. Both NC and CC interactions induced by electron-neutrinos are classified as NC events, except for those where a hadron gives rise to a "long" shower. The absolute numbers of these backgrounds were derived from the known kaon content in the parent meson beam.

The corrections discussed are given, for both experiments, in Table 4.9. Some other minor corrections related to various efficiencies involved in the data acquisition and event selection have been omitted. The final corrected event numbers for CDHS and CHARM are also given in the table. These event numbers translate into the following values for the ratio R_ν:

$$R_\nu = \begin{cases} 0.3072 \pm 0.0025 \text{ (stat.)} \pm 0.0020 \text{ (syst.)} & \text{CDHS} \\ 0.3098 \pm 0.0029 \text{ (stat.)} \pm 0.0011 \text{ (syst.)} & \text{CHARM} \end{cases}$$

Measurements of $r = \sigma_{\bar{\nu}}^{CC}/\sigma_{\nu}^{CC}$ from an exposure to antineutrinos yield the results

$$r = \begin{cases} 0.390 \pm 0.010 & \text{CDHS} \\ 0.456 \pm 0.012 & \text{CHARM} \end{cases}$$

In order to determine $\sin^2 \theta_W$ from R_ν by virtue of (4.37) one has to correct for the physical idealisations used in the derivation of this relation. These corrections were calculated in the framework of a QCD quark parton model of the nucleon. Both experiments have followed very similar procedures in their model calculations so that the resulting values for $\sin^2 \theta_W$ are directly comparable. The effects of the various corrections applied by the two experiments and their errors are shown in Table 4.10. The errors of the corrections in essence result from the uncertainty in the experimental input to the quark parton model calculations, such as the CKM matrix elements, the size of the sea contributions from s and c quarks etc.. A few corrections are specific to the individual experiment, such as the correction for non-isoscalar targets by CDHS using an iron target (in contrast, the marble absorber of CHARM has a ratio of u/d valence quarks very close to one), or the muon mass correction term by CHARM who rely on muon identification to seperate CC from NC events and thus keep track of the small phase space suppression in CC events. The source of the other common corrections have been discussed above, except those for the longitudinal structure function and the radiative correction. The former one takes account of the fact that the Callan-Gross relation, a consequence of the simple quark parton model, is not exactly fulfilled due to QCD corrections to the QPM. The radiative corrections were calculated with $\sin^2 \theta_W$ expressed in the on-shell renormalisation scheme according to Wheater and Llewellyn-Smith (1982) in the case of CDHS, and according to Bardin et al. (1986) for CHARM. These corrections are also given in Table 4.10. Including all corrections except the one for electroweak radiative effects the values for $\sin^2 \theta_W$ are given as

$$\sin^2 \theta_W = \begin{cases} 0.236 \pm 0.006 & \text{CDHS} \\ 0.245 \pm 0.006 & \text{CHARM} \end{cases} \tag{4.39}$$

where the experimental and theoretical errors, excluding the error due to the charm mass, have been added in quadrature. The largest theoretical uncertainty is in the choice of the charmed quark mass m_c which affects the threshold suppression of charm production. Both experiments therefore have chosen to present the result for $\sin^2 \theta_W$ as a linear function of m_c, with $m_c = 1.5$ GeV/c^2 as the

Table 4.10. Corrections to $\sin^2 \theta_{\mathrm{W}}$ and their uncertainties as calculated using a QCD quark parton model for the nucleon. Note that the data from CDHS and CHARM have been obtained with $E_{\mathrm{cut}}^h = 10$ GeV and 4 GeV, respectively

| Source | $\Delta \sin^2 \theta_{\mathrm{W}} \pm \delta \sin^2 \theta_{\mathrm{W}}$ | |
	CDHS	CHARM
Non-isoscalar target	-0.0090 ± 0.0009	
Non-strange sea	$+0.0022 \pm 0.0003$	
Long. struct. function F_{L}	$+0.0006 \pm 0.0006$	
W^2 thresholds, F_{L}		$+0.0005 \pm 0.0005$
Quark mixing	$+0.0031 \pm 0.0003$	0.0000 ± 0.0010
Strange sea	$+0.0043 \pm 0.0010$	-0.0074 ± 0.0010
Charm sea	$+0.0003 \pm 0.0003$	$+0.0015 \pm 0.0010$
Charm mass	$+0.0100 \pm 0.0040$	$+0.0140 \pm 0.0040$
Muon mass		$+0.0011 \pm 0.0001$
Radiative correction	-0.0110 ± 0.0020	-0.0092 ± 0.0020
Total ($m_c = 1.5$ GeV/c^2)	$+0.0005 \pm 0.0030$	$+0.0005 \pm 0.0030$

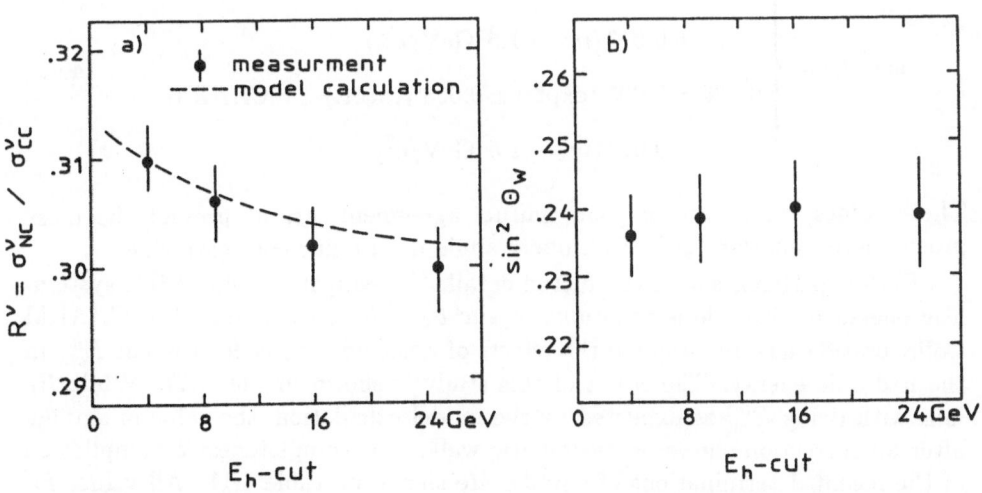

Fig. 4.17. Dependence of R_ν and $\sin^2 \theta_{\mathrm{W}}$ on the cut in the hadronic energy (CHARM collaboration)

Table 4.11. Recent measurements of $\sin^2\theta_W$ in neutrino nucleon scattering. All results are obtained with the condition $\varrho = 1$. The values given for $\sin^2\theta_W$ are corrected for electroweak radiative effects

Experiment	Method	$\sin^2\theta_W$
BEBC Jones et al. '86	$\sigma_p^{NC}/\sigma_p^{CC}$	0.225 ± 0.030
CCFRR Reutens et al. '85	$(\sigma_\nu^{NC} - \sigma_{\bar\nu}^{NC})/(\sigma_\nu^{CC} - \sigma_{\bar\nu}^{CC})$	$0.242 \pm 0.011 \pm 0.005$
FMM Bogert et al. '85	$\sigma_\nu^{NC}/\sigma_\nu^{CC}$, $\sigma_{\bar\nu}^{NC}/\sigma_\nu^{NC}$	$0.246 \pm 0.012 \pm 0.013$
CDHS Bergsma et al. '86	$\sigma_\nu^{NC}/\sigma_\nu^{CC}$	$0.225 \pm 0.005 \pm 0.005$
CHARM Abramovicz et al. '86	$\sigma_\nu^{NC}/\sigma_\nu^{CC}$	$0.236 \pm 0.005 \pm 0.005$
BNL (USA-Japan) Abe et al. '86	$\dfrac{\sigma(\nu p \to \nu p)}{\sigma(\bar\nu p \to \bar\nu p)}$	$0.220 \pm 0.016^{+0.023}_{-0.031}$
Average	all data	$0.233 \pm 0.003 \pm 0.005$

central value. The linear approximation is valid for m_c between 1 and 2 GeV/c^2. Including also the electroweak radiative correction the final answer to $\sin^2\theta_W$ from both experiments is

$$
\sin^2\theta_W = \begin{cases}
0.225 \pm 0.005\,(\text{expt.}) \pm 0.003\,(\text{theor.}) & \text{CDHS} \\[4pt]
\quad + 0.013\,(m_c - 1.5\,[\text{GeV}/c^2]) & \\[6pt]
0.236 \pm 0.005\,(\text{expt.}) \pm 0.003\,(\text{theor.}) & \text{CHARM} \\[4pt]
\quad + 0.012\,(m_c - 1.5\,[\text{GeV}/c^2])\,.
\end{cases}
\tag{4.40}
$$

These values, which are in good mutual agreement, are at present the most precise determinations of the Weinberg angle from ν nucleon scattering.

Both experiments have conducted detailed investigations of possible systematic effects in their determinations of $\sin^2\theta_W$. As an example, the CHARM Collaboration has investigated the effects of a specific choice for the cut E_{cut}^h in the hadronic energy. The result of this study is shown in Fig. 4.17. While R_ν falls with rising E_{cut}^h as expected by the model calculation, the value of $\sin^2\theta_W$ after all correction shows no systematic walk. For completeness a compilation of the recent determinations of $\sin^2\theta_W$ are shown in Table 4.11. All values for $\sin^2\theta_W$, except BEBC, are corrected for electroweak radiative effects. Within the errors given the various determinations are in very good agreement.

As can be seen from (4.40) there seems to be no point in further decreasing the experimental error as long as the theoretical uncertainties concerning the various corrections listed in Table 4.10 and, in particular, the proper treatment of the charm suppression, have not been substantially reduced. Taking the weighted average of all deep-inelastic neutrino scattering experiments listed in Table 4.11 one obtains

$$\sin^2 \theta_W = 0.233 \pm 0.006$$

which is a substantial improvement in accuracy over the results from e^+e^- experiments and neutrino electron scattering presented earlier. Remarkably, all values, although derived under quite different physical circumstances (scattering partners, observables, experimental conditions) are in excellent agreement with each other.

5. Charged-Lepton Quark Scattering

In the historical development of understanding and determining the structure of the weak neutral current the scattering of charged leptons on nuclei has played a key role. In particular, the polarised-electron deuteron scattering experiment at SLAC in 1979 has paved the way for the standard theory, disproving a number of competing models to describe the weak neutral current (the famous "SLAC masacre"). This and other experiments using high energy polarised charged leptons and experiments on low energy atomic spectroscopy have shown that the electron (muon) and the nucleon, or quark in more basic terms, couple through a weak neutral current in addition to their electromagnetic interactions precisely in the way predicted by the standard theory. Measurement of electroweak interference effects in various charged-lepton quark scattering processes provide powerful consistency checks of the standard theory since the weak neutral couplings of the participating fermions have been determined quite accurately in complementary reactions such as $\nu q, \nu e$, and $e^+ e^-$ scattering, as discussed in the previous chapters.

5.1 Inclusive Scattering of Polarised Electrons

The scattering of electrons with the charged constituents of the nucleon are dominated at presently accessible energies by the electromagnetic interaction, which is mediated in lowest order by one-photon exchange in the t channel. Weak neutral current contributions to the scattering amplitude are introduced by adding Z^0 exchange, similar to $e^+ e^-$ interactions. Due to the low momentum transfers involved one can work in the local limit (3.11), where weak effects occur with a strength characterised by the Fermi coupling constant G. The total squared amplitude contains, in close analogy to the Bhabha amplitudes in the t channel (see Table 3.2), an electromagnetic term proportional to α^2, an electroweak interference term proportional to αG, and a purely weak term proportional to G^2 which can be safely neglected. The electroweak interference consists of a scalar portion (terms proportional to $v_e v_q$ or $a_e a_q$) which conserves parity giving rise to charge asymmetries as discussed for $e^+ e^-$, and a pseudoscalar portion violating parity (terms proportional to $v_e a_q$ and $a_e v_q$) which is responsible for polarisation effects (see also (3.29)). The sign of the pseudoscalar term depends on the handedness of the system under investigation. Measuring a difference in rates with incident electrons of opposite helicities will therefore reveal weak neutral currents in eq scattering.

The polarisation asymmetry $A(y)$ is defined as

$$A(y) = \frac{d\sigma_R(y) - d\sigma_L(y)}{d\sigma_R(y) + d\sigma_L(y)}, \tag{5.1}$$

where $y = (E - E')/E$ is calculated from the energies of the initial (E) and final (E') electron in the laboratory system. Within the simple quark-parton model the cross sections $d\sigma$ can formally be obtained from the Bhabha scattering t-channel amplitudes in Table 3.2, where the chiral couplings and the charge of the e^+ are replaced by the corresponding quantities for the u and d quarks. Furthermore, using the replacement $(1 - y)$ for $(1 + \cos\theta)/2$ (see also Sect. 4.2) the cross section for right- and left-handed electrons can be written as

$$d\sigma_R(y) \propto \sum_q \int \left(|\varepsilon_{RR}^q|^2 + |\varepsilon_{RL}^q|^2 (1 - y)^2 \right) x\, q_q(x)\, dx$$

$$d\sigma_L(y) \propto \sum_q \int \left(|\varepsilon_{LR}^q|^2 + (1 - y)^2 |\varepsilon_{LL}^q|^2 \right) x\, q_q(x)\, dx\,,$$

where x denotes the fraction of momentum of the nucleon carried by the quark q. As a reminder, the index convention for ε_{kl} is such that k denotes the handedness of the electron and l the handedness of the quark. Neglecting sea quark contributions, one obtains for $A(y)$ for an isoscalar target (see also Cahn and Gilman, 1978):

$$A(y) = \frac{\left(\sum_q |\varepsilon_{RR}^q|^2 - \sum_q |\varepsilon_{LL}^q|^2\right) + \left(\sum_q |\varepsilon_{RL}^q|^2 - \sum_q |\varepsilon_{LR}^q|^2\right)(1 - y)^2}{\left(\sum_q |\varepsilon_{RR}^q|^2 + \sum_q |\varepsilon_{LL}^q|^2\right) + \left(\sum_q |\varepsilon_{RL}^q|^2 + \sum_q |\varepsilon_{LR}^q|^2\right)(1 - y)^2}. \tag{5.2}$$

The sums run over u and d quarks, with distribution functions $q_q(x)$ of the valence quarks already encountered in deep-inelastic νq scattering. For an isoscalar target $q_u(x) = q_d(x)$ so that the x dependence of A drops out. Neglecting the small purely weak part, the functions $|\varepsilon_{kl}^q|^2$ can be expressed in terms of the vector and axial-vector constants v_q, a_q, the charge Q_q of the target quark q and the momentum transfer squared t as follows:

$$|\varepsilon_{LL}^q|^2 = Q_q^2 - Q_q(v_e v_q + a_e a_q + v_e a_q + a_e v_q)\mathrm{Re}\{g(t)\}$$

$$|\varepsilon_{LR}^q|^2 = Q_q^2 - Q_q(v_e v_q - a_e a_q - v_e a_q + a_e v_q)\mathrm{Re}\{g(t)\}$$

$$|\varepsilon_{RL}^q|^2 = Q_q^2 - Q_q(v_e v_q - a_e a_q + v_e a_q - a_e v_q)\mathrm{Re}\{g(t)\} \tag{5.3}$$

$$|\varepsilon_{RR}^q|^2 = Q_q^2 - Q_q(v_e v_q + a_e a_q - v_e a_q - a_e v_q)\mathrm{Re}\{g(t)\}\,.$$

In the denominator of (5.2) the electroweak interference term can safely be neglected compared to the electromagnetic one, which is given by the square of the quark charge. In the local limit for the propagator term $g(t)$ (see (3.11)), using the positive quantity $Q^2 = -t$, the asymmetry is finally given by

$$A(y) = -\frac{G}{\sqrt{2}}\frac{Q^2}{4\pi\alpha}\left(\frac{3}{5}\right)\left[a_1 + a_2\frac{1-(1-y)^2}{1+(1-y)^2}\right]$$ (5.4)

with

$$a_1 = -a_e\left(2v_u - v_d\right)$$

$$a_2 = -v_e\left(2a_u - a_d\right).$$ (5.5)

In the standard theory a very weak y dependence is expected since a_2 is proportional to the small vector coupling constant of the electron. The vector couplings of the quarks, on the other hand, are sizeable and opposite in sign (see Table 2.2) so that the coefficient a_1 is of order 1. The measured quantity A_{exp} is finally related to (5.4) by the degree of polarisation $P(e)$ of the electron beam

$$A_{\mathrm{exp}}(y) = P(e)\,A(y).$$ (5.6)

As shown in Chap. 3, the size of the electroweak interference effect is of order Q^2/M_Z^2. For eq scattering with typical $Q^2 \sim [1\,\mathrm{GeV}]^2$ one therefore expects a polarisation asymmetry $A(y)$ in the scattering cross section of order 10^{-4}, which has indeed been observed in the inclusive scattering of longitudinally polarised electrons on an isoscalar target (deuterium) at SLAC (Prescott et al., 1978, 1979). In the experiment, $P(e)$ was 37 %. This experiment helped to exclude a number of models which had been formulated in addition to the GSW model at times where confusing data were avalaible from atomic physics experiments. The SLAC measurements for $A(y)$ are shown in Fig. 5.1 together with predictions from the standard theory and other electroweak gauge models competing at that time (Bilenky et al., 1977; Cheng and Li, 1977).

In order to extract the coefficients a_1 and a_2 from (5.4) the data have to be corrected for electromagnetic radiative effects (Mo and Tsai, 1978). Radiative corrections do not generate polarisation asymmetries but change the kinematic quantities Q^2 and y which are calculated from the nominal beam energy and the spectrometer setting. The corrected results for a_1 and a_2 are (Prescott et al., 1979):

$$a_1 = 1.80 \pm 0.48$$

$$a_2 = 0.91 \pm 1.50.$$ (5.7)

These values can be compared to the standard theory prediction, using Table 2.2 and $\sin^2\theta_{\mathrm{W}} = 0.23$,

$$a_1(\mathrm{GSW}) = 1.47$$

$$a_2(\mathrm{GSW}) = 0.24\,,$$

and are found in good agreement with the expectations.

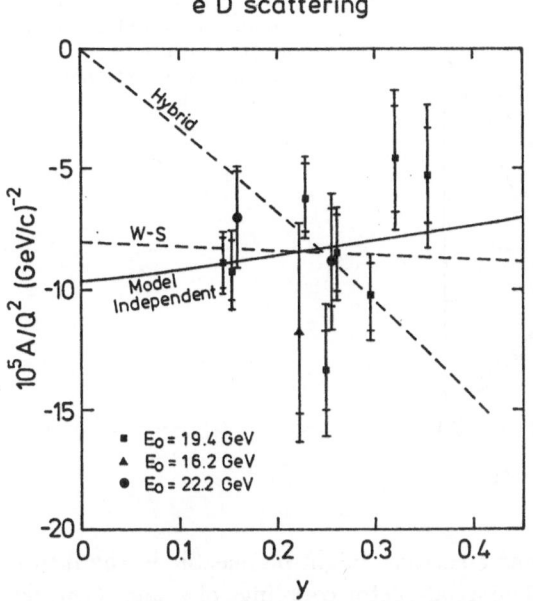

e D scattering

Fig. 5.1. Measurements of the polarisation asymmetry $A(y)$ in eD scattering (Prescott et al., 1979). Also shown are the predictions of the standard theory and another competing electroweak gauge model

The measurement for the coefficients a_1 and a_2 can be used to put limits on the allowed values for the vector coupling constants of u and d quarks. Figure 5.2 shows the 90 % C.L. regions in the $v_u - v_d$ plane allowed by the SLAC eD experiment together with the expectation from the standard theory, where $a_e = -1$ has been assumed. Also shown in the figure are the 90% C.L. boundaries from atomic parity violation experiments, to be discussed in the next section.

More generally, these contours can be considered as limits for the couplings C_{1u} and C_{2u} which have been introduced by Hung and Sakurai (1977) on the basis of a very general Lorentz structure of the weak neutral current. In terms of the weak coupling constants v_i and a_i, the Hung-Sakurai couplings are given as

$$C_{1u} = \frac{1}{4}\, a_e\, v_u$$

$$C_{1d} = \frac{1}{4}\, a_e\, v_d\,. \tag{5.8}$$

One notices that factorisation is not assumed when the couplings C_{1u} and C_{2u} are used. The limits for these constants are obtained from Fig. 5.2 by rescaling the axes according to (5.8). A detailed discussion on the SLAC experiment and its relevance for constraining weak neutral current parameters has been published by Commins and Bucksbaum (1980).

In principle, the measurement of elastic eN scattering and Δ production ($ep \to e\Delta$) using polarised e beams would help to constrain the axial couplings of the quarks (Cahn and Gilman, 1978, Gilman and Tsao, 1979). No data are available for these reaction yet, although interesting medium energy experiments,

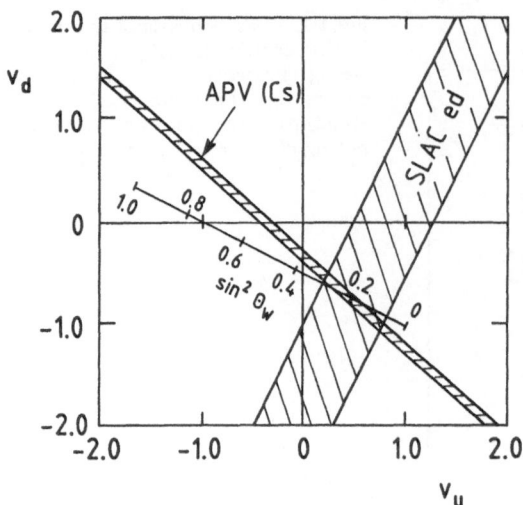

Fig. 5.2. 90% C.L. contours in the $v_u - v_d$ plane from polarised eD scattering (Prescott et al., 1979) and atomic parity violation in Cs (Bouchiat et al., 1983)

in particular Δ excitation by polarized electrons, might be feasible in the future (see, e.g., Achenbach et al., 1986). The axial-vector couplings of u and d quarks thus remain poorly determined from eD scattering owing to the very weak y dependence of the polarisation asymmetry in the vicinity of $\sin^2 \theta_W = 0.25$.

The SLAC data have also been used to extract the value of $\sin^2 \theta_W$. With the GSW ansatz for the quark and electron couplings in (5.4), Prescott et al. (1979) have carried out fits to the coefficients a_1 and a_2 with the weak mixing angle as the only free parameter and obtain:

$$\sin^2 \theta_W = 0.224 \pm 0.012 \pm 0.008 . \qquad (5.9)$$

This result has been shown to depend rather weakly on the specific assumptions of the quark parton model which was used to interpret the data, as long as $\sin^2 \theta_W$ is in the vicinity of 1/4 (Fritzsch, 1979; Bjorken, 1978a; Wolfenstein, 1978). Also, as shown by Fajfer and Oakes (1984), higher twist effects will not contribute significantly. They estimate a correction to $\sin^2 \theta_W$ due to higher twist effects of less than 1%.

The value for $\sin^2 \theta_W$ given in (5.9) is not corrected for electroweak radiative effects. These corrections have been carried out in the on-shell renormalisation scheme and recent calculations by Marciano (1985) lead to the following radiatively corrected value of $\sin^2 \theta_W$ from polarised eD scattering

$$\sin^2 \theta_W = 0.218 \pm 0.012 \pm 0.008 .$$

5.2 Atomic Parity Violation

Parity violation experiments in atoms appear complementary in many ways to deep inelastic electron nucleon scattering. In both cases the reaction partners are the same, electrons are scattered on quarks. However, most evidently, the scales of momentum transfers involved are very different, with about 1 GeV/c for $e\,D$ scattering and 1 to 10 MeV/c for heavy atoms (the surprisingly large value of Q^2 for an atomic transition is related to the short range of the weak interaction: the valence electron moves in an unscreened Coulomb field generated by the nucleus of charge Ze, see Bouchiat and Piketty, 1983). Secondly, in deep inelastic scattering the nucleus is broken up and the quarks interact incoherently with the probing current, whereas in atomic systems, due to the very small momentum transfers involved, the nucleus stays intact and the quarks act coherently. Furthermore, as will be shown below, the combinations of weak couplings involved in atomic physics experiments are almost orthogonal to the ones encountered in deep inelastic electron nucleon scattering.

Similar to eN scattering the parity-violating effects in atoms arise from pseudoscalar terms proportional to $a_e v_A$ and $v_e a_A$, where A stands for the nucleus investigated. Since the atomic electrons interact with the nucleus as a whole, the weak charges v_A and a_A of the nucleus are obtained by coherently adding the respective quark contributions for Z protons and N neutrons:

$$v_A = (2\,v_u + v_d)\,Z + (v_u + 2\,v_d)\,N$$

$$a_A = (2\,a_u + a_d)\,Z + (a_u + 2\,a_d)\,N\,. \tag{5.10}$$

Using the standard theory (see Table 2.1), these relation can be expressed in terms of $\sin^2\theta_W$:

$$v_A = (1 - 4\sin^2\theta_W)\,Z - N$$

$$a_A = Z - N\,. \tag{5.11}$$

For heavy nuclei the vector coupling v_A, also called the weak charge Q_W in the literature,

$$Q_W \equiv v_A$$

is expected to be large, of order N, since $\sin^2\theta_W$ is close to 1/4. In addition, the weak charge is multiplied by a_e in the parity-violating terms. In contrast, the small term a_A will be further suppressed by v_e, entailing $v_e a_A \approx 0$. This is equivalent to the statement that the axial nucleonic current, in the limit of zero momentum transfer, is proportional to the nucleon spins, which tend to cancel in pairs in the sum over nucleons. Thus a favourable situation remains where the atomic parity violation effects are proportional to the product $a_e v_A$ with both factors large.

As for the other observables generated through electroweak interference, the parity-violating effects are naively expected to be of order Q^2/M_Z^2, which is exceedingly small for the momentum transfers $Q^2 \sim \alpha^2 \, m_e^2$ involved in atomic transitions. One thus would predict typical asymmetries around 10^{-15}, which is clearly unmeasurable. The effects, however, can be considerably enhanced, as noticed first by Bouchiat and Bouchiat (1974), when heavy atoms are used. In heavy atoms an enhancement factor of order Z^3 becomes operative, where one power of Z arises from Q_W and the two others from the point-like structure of the interaction. With an additional relativistic enhancement factor of about $3 - 10$, the final enhancement is of order $10^6 - 10^7$ for heavy atoms with Z between 50 and 80.

In atomic parity violation experiments one observes an optical transition between a pair of states with the same nominal parity (magnetic dipole transition). If only the electromagnetic interaction is considered, the wave functions associated with atomic energy levels are pure eigenstates of parity. The weak neutral current, interfering with the electromagnetic current, violates parity and consequently each level will receive small admixtures of the opposite parity state. Thus a magnetic dipole transition, described by the (real) matrix element M_1, will receive a parity-violating electric dipole piece E_1^{pv}, implying a total transition probability

$$W \sim M_1^2 + |E_1^{\mathrm{pv}}|^2 \pm 2 \operatorname{Im}\{E_1^{\mathrm{pv}}\} \, M_1 \,, \qquad (5.12)$$

where, similar to polarised eD scattering, the sign in front of the interference term will depend on the handedness of the system, in this case the helicity of the absorbed (or emitted) photon. The relative size of the parity-violating effect can be obtained from (5.12) and is proportional to $\operatorname{Im}\{E_1^{\mathrm{pv}}\}/M_1$.

In a medium of atomic vapor of density D, the transition probability can be translated to an index of refraction for circularly polarised photons with frequency ω, which assumes the following form in the vicinity of a resonant frequency ω_0 (Commins and Bucksbaum, 1980):

$$n_\pm = 1 - \frac{2\pi D}{\hbar} \, \langle M_1^2 \pm 2 \operatorname{Im}\{E_1^{\mathrm{pv}}\} \, M_1 \rangle f(\omega, \omega_0) \,. \qquad (5.13)$$

The $\pm$ sign corresponds to the two helicity states of the incoming photon. The complex function $f(\omega, \omega_0)$ describes the line shape of the resonance. The matrix element is averaged over atomic polarisations in the initial state and summed over the possible final states. The complex index of refraction gives rise to two parity-violating phenomena:

Optical rotation : A difference in the real part of the $n_\pm$ indices for the two helicity states implies different velocities for right or left circularly polarised light in the medium. Thus a *linearly* polarised laser beam tuned to the resonance frequency ω_0 will suffer a rotation ϕ of the plane of polarisation when propagating a unit distance through the vapor. Since ϕ is proportional to the difference between the real parts of the refractive indices n_L and n_R for left and right circularly polarised light in the médium, the size of ϕ is given by (see (5.13))

$$\phi \sim 2\text{Im}\{E_1^{\text{pv}}\}\cdot M_1 . \tag{5.14}$$

Optical rotation angles $\phi \sim 10^{-8} - 10^{-7}$ radians/absorption length can be expected for heavy atoms such as bismuth (Bi) with the above Z^3 enhancement.

Circular dichroism : A difference in the imaginary part of $n_\pm$ for the two helicity states implies different amounts of absorption for right or left circularly polarised light in the medium. Thus the amount of fluorescence light will depend on the handedness of the circularly polarised laser light. Using (5.12), the resulting cross section asymmetry is given by

$$\Delta = \frac{\sigma_+ - \sigma_-}{\sigma_+ + \sigma_-} = \frac{2\,\text{Im}\{E_1^{\text{pv}}\}\,M_1}{M_1^2 + |\{E_1^{\text{pv}}\}|^2} \sim \frac{2\text{Im}\{E_1^{\text{pv}}\}}{M_1} . \tag{5.15}$$

This type of experiment was originally proposed by Bouchiat and Bouchiat (1974).

For experiments on circular dichroism one may exploit the relation $\Delta \sim \text{Im}\,E_1^{\text{pv}}/M_1$ and further enhance the parity-violating effect by choosing strongly forbidden M_1 transitions. With this additional enhancement, dichroic asymmetries $\Delta \sim 10^{-4} - 10^{-3}$ can be expected in the standard theory.

One of the problems in atomic physics experiments is the interpretation of the data in a specific theoretical framework such as the standard theory. Precise tests with heavy atoms were notoriously plagued with the uncertainties of the rather complex atomic physics calculations. Among others, many body problems, screening effects, and incompletely known potentials have to be mastered. Calculations for various promising atoms have been done by a number of authors (see, e.g., Novikov et al., 1976; Carter and Kelly, 1979; Sandars, 1980; Martenssen et al., 1981 for calculations on bismuth), with sometimes considerable spread in the predictions.

Also the experimental results on atomic parity violation have been contradictory in the past. Most experiments have investigated the optical rotation technique in atomic systems such as bismuth (Bi), thallium (Tl) or lead (Pb). While some older experiments reported null results (Baird et al., 1977; Apperson, 1979; Bogdanov et al., 1980), the present wealth of data has established a clear parity-violating effect in atomic physics. Experiments on the Bi 648 μm line have been performed in Novosibirsk (Barkov and Zolotorev, 1978), the Seattle group has investigated Bi (876 μm, Hollister et al., 1981) and lead (Emmons et al., 1983). Other groups in Oxford and Moscow have communicated their results on the Bi 648 μm line (see Piketty, 1984).

The optical rotation measurements are shown in Table 5.1. Also given are the theoretical expectations for the various experiments, assuming the standard theory with $\sin^2\theta_W = 0.23$. Errors have been attached to the theoretical estimates which are indicative of the spread of calculations from different authors. Details on experimental realisations and the theoretical interpretation of results,

Table 5.1. Results from atomic parity violation experiments using the optical rotation technique. Also shown are the theoretical predictions, assuming the standard theory with $\sin^2 \theta_W = 0.23$. Radiative corrections have been applied to the predictions (see text)

Experiment	$10^8 \cdot \mathrm{Im}\{E_1^{\mathrm{pv}}\}/M_1$	Theory
Bismuth , $\lambda = 648\ \mu$m		
Novosibirsk , 1978	-20.2 ± 2.7	
Oxford , 1984	$-\ 9.3 \pm 1.5$	-13.0 ± 3.0
Moscow , 1984	$-\ 7.8 \pm 1.8$	
Bismuth , $\lambda = 876\ \mu$m		
Seattle , 1981	-10.4 ± 1.7	-10.0 ± 2.5
Lead , $\lambda = 1280\ \mu$m		
Seattle , 1983	$-\ 9.9 \pm 2.5$	-12.5 ± 2.0

especially the involved spectroscopic calculations, can be found, e.g., in the review articles of Commins and Bucksbaum (1980), Fortson and Wilets (1980), and Barkov (1982).

In the second generation of atomic physics experiments, circular dichroism has been investigated by the groups at Paris (Bouchiat et al., 1982, 1984) and Boulder (Gilbert et al., 1985), using cesium, and at Berkeley (Buchsbaum et al., 1981; Drell and Commins, 1984) using thallium. In these experiments an external elctrostatic field, perpendicular to the laser beam, is used to induce electric dipole (E_1) transitions, rather than M_1 as in the optical rotation experiments: The interference of the amplitude E_1^{pv} with the E_1 amplitude results in a parity-violation electronic polarisation (PVEP) of the laser-excited atom. The PVEP is perpendicular to the electric field and to the incoming laser light, and has specific symmetry properties discriminating it from other parity-conserving polarisations (PCEP). As an example, the PVEP maintains its direction when both the helicity and the direction of the laser light are reversed, i.e. when the laser light is mirror-reflected. In contrast, the part of the PCEP which is parallel to the PVEP changes sign when the laser light is mirror-reflected. This property can be advantageously used to pass the laser light many times through the cell containing the atomic vapour, each time reflected by spherical mirrors at either end of the cell. In the ideal case, the PCEP has been reduced to zero, and the circular polarisation of the decay fluorescence light, which one finally observes, is only due to the PVEP. This simplified discussion can only cover the most simple principles of the ingenious atomic parity violation experiments using Stark-induced transitions. Some of the above cited experiments have used more complex arrangements with crossed electric and magnetic fields for light trans-

Table 5.2. Results from atomic parity violation experiments using the circular dichroism technique. Also shown are the theoretical predictions, assuming the standard theory with $\sin^2\theta_W = 0.23$. Radiative corrections have been applied to the predictions (see text)

Experiment	$\mathrm{Im}\{E_1^{\mathrm{pv}}\}/\beta[\,\mathrm{mV}]$	Theory
Thallium		
Berkeley , 1981	$-1.80 \pm 0.45\,^{+0.20}_{-0.15}$	-1.15 ± 0.35
Berkeley , 1984	$-1.73 \pm 0.26 \pm 0.07$	
Cesium		
Paris , 1982	$-1.34 \pm 0.22 \pm 0.11$	
Paris , 1984	$-1.78 \pm 0.26 \pm 0.12$	-1.67 ± 0.13
Boulder, 1984	-1.63 ± 0.13	

mission measurements, employing atomic beams rather than vapour to reduce backgrounds from atomic collisions, molecular effects etc..

The above experiments basically yield the quantity $\mathrm{Im}\{E_1^{\mathrm{pv}}\}/\beta$, where β is the vector polarisability of the atom under study. With the help of atomic model calculations both E_1^{pv} and β can be calculated, their ratio depending on the vector charge v_A of the nucleus. Several spectroscopic calculations for cesium exist (see, e.g., Bouchiat et al., 1983; Martenssen-Pendrill, 1985), which agree within about 10 %. Taking an average over these calculations, including an error reflecting the theoretical uncertainty, leads to the following relation between the measurement and the weak charge Q_W:

$$\text{Cesium}:\quad \mathrm{Im}\{E_1^{\mathrm{pv}}\}/\beta = -(1.81 \pm 0.09)\,Q_W/(-N)\,[\mathrm{mV/cm}]\,. \tag{5.16}$$

Measurements of the induced electric dipole moment $\mathrm{Im}\{E_1^{\mathrm{pv}}\}/\beta$ for cesium by the Paris and Boulder groups, and for thallium by the Berkeley group, are given in Table 5.2. From the experimental value of the induced electric dipole moment one extracts, by virtue of (5.16), the quantity Q_W which can then be compared with the expectation (5.10). A relation equivalent to (5.16) exists also for thallium. However, the theoretical uncertainties are much larger in this case due to the more complex electronic shell configuration. This uncertainty may be reflected in the large discrepancy between the measurement and the theoretical estimate (see Table 5.2). It should be stressed here that the atomic physics calculations for cesium are the most reliable due to the fact that cesium has only one electron in the outermost shell while the situation is more complicated for thallium (3 electrons), lead (4 electrons), and bismuth (5 electrons).

Taking the average over the three Cs measurements, one finds the following value for the weak vector charge Q_W:

$$Q_W = -69.0 \pm 4.8\,(\text{stat.}) \pm 3.4\,(\text{syst.}) \pm 3.5\,(\text{atom.phys.}) \ . \tag{5.17}$$

Before comparing this value of Q_W with (5.11) one has to consider the effect of radiative corrections, for which calculations exist (Marciano and Sirlin, 1983). The calculations are carried out in the framework of the standard theory using the modified minimal substraction ($\overline{\text{MS}}$) scheme and provide corrections to the strength of the weak neutral current, characterised by the ϱ parameter, and to the Weinberg angle. More specifically, equation (5.11) is modified to

$$v_A = \varrho_{\text{pv}}\left[(1 - 4\,\kappa_{\text{pv}}\,\sin^2\hat{\theta}_W\,(M_W)\,Z - N\right]$$

$$a_A = \varrho_{\text{pv}}\left[Z - N\right] \ . \tag{5.18}$$

In absence of radiative corrections $\varrho_{\text{pv}} = \kappa_{\text{pv}} = 1$. To $O(\alpha)$, the coefficients ϱ_{pv} and κ_{pv} receive corrections depending on the renormalised mixing angle $\sin^2\hat{\theta}_W\,(M_W)$, the masses m_t and m_ϕ of the top quark and the Higgs boson, and M_W. The result of the calculation by Marciano and Sirlin (1983) using $\sin^2\hat{\theta}_W\,(M_W) = 0.215$, $m_t = 20$ GeV, $m_\phi = m_Z$ and $m_W = 83$ GeV is

$$\varrho_{\text{pv}} = 0.973, \qquad \kappa_{\text{pv}} = 1.003 \ .$$

For the case of cesium with $Z = 55$, $N = 78$ the standard theory prediction, according to (5.18), is $Q_W = -71.8$, in very good agreement with the average from the three experiments. As can be seen from (5.11), the atomic physics experiments are not well suited to determine the Weinberg angle for values of $\sin^2\theta_W$ in the vicinity of $1/4$ due to cancellation effects. From the Cs average one obtains

$$\sin^2\theta_W = 0.217 \pm 0.020 \pm 0.016 \pm 0.016 \ ,$$

where the errors are related to statistics, systematics, and uncertainties in the atomic physics calculations.

With the experimental value of Q_W given in (5.17) one can derive 90% C.L. limits for the vector couplings of u and d, linked through relation (5.10)

$$188\,v_u + 211\,v_d = -69.0 \pm 6.4 \ , \tag{5.19}$$

where all errors have been added in quadrature. The corresponding contour is drawn in Fig. 5.2 together with the corresponding limits from eD scattering (see (5.5)), assuming $a_e = -1$. It is evident from the figure that the combination of couplings from atomic parity-violation is almost orthogonal to the one from eD scattering.

The analysis of atomic parity-violation in heavy atoms is based on calculations of complex electron configurations with their inherent theoretical uncertainties. Even without these uncertainties experiments with heavy atoms could only measure the vector coupling constants v_A, since a_A is multiplied by the very small electronic vector charge v_e. This has motivated the investigation of parity-violating effects in hydrogen, deuterium and tritium. The effects will be much smaller due to the missing enhancement factors and no results have been published yet. A discussion of these interesting experiments, and the techniques and problems involved, can be found, e.g., in Commins and Bucksbaum (1980), Commins (1981) and Fortson and Lewis (1984).

5.3 Muon-Quark Scattering

Data on deep-inelastic scattering of longitudinally polarised μ^+ and μ^- on an isoscalar target (carbon) have been reported by the Bologna-CERN-Dubna-München-Saclay Collaboration (Argento et al., 1983) which clearly exhibit the effects of weak neutral currents in the muon-quark interaction. The physics of μC scattering is quite different from the SLAC eD experiment: The Q^2 range studied in μC is about two orders of magnitude larger compared to eD (a $\mu^\pm$ beam of 120 and 200 GeV from the SPS was used), so that the electroweak asymmetry, according to the naive expectation, is as large as 1 %. This asymmetry, in contrast to the SLAC experiment, is mainly parity-conserving (see below). Furthermore, the use of muon beams provides a powerful tool in constraining the weak coupling constants v_μ and a_μ of the muon, of which so far only the axial-vector constant a_μ has been measured (see Chapter 3). This is due to the fact that the polarised muons are obtained from π decays, leading to preferentially left-handed positive muons and right-handed negative muons. Finally, the resulting cross section asymmetry constrains the axial-vector coupling constants of the quarks, rather than the vector couplings as in the SLAC experiment. Further details of the pertinent physics as well as a description of the beams, set-up, and systematic studies of the μC experiment can be found in Klein (1985).

The quantity measured is the cross section asymmetry $B(y)$, defined as

$$B(y) = \frac{d\sigma\,(\mu_L^+) - d\sigma\,(\mu_R^-)}{d\sigma\,(\mu_L^+) + d\sigma\,(\mu_R^-)} \ . \tag{5.20}$$

Writing down the expression for the scattering cross section of left-handed μ^+ and right-handed μ^-

$$d\sigma(\mu_L^+) = \sum_q \int \left(|\bar{\varepsilon}_{LR}|^2\,(1-y)^2 + |\bar{\varepsilon}_{LL}|^2 \right) x\,q(x)\,dx$$

$$d\sigma(\mu_R^-) = \sum_q \int \left(|\varepsilon_{RR}|^2 + |\varepsilon_{RL}|^2\,(1-y)^2 \right) x\,q(x)\,dx \ ,$$

the asymmetry $B(y)$ can be calculated, assuming the simple quark model, in complete analogy to electron-quark scattering, using the coefficients ε_{kl}^i for μ^- and $\bar{\varepsilon}_{kl}^i$ for μ^+:

$$B(y) = \frac{\left(\sum_q |\bar{\varepsilon}_{LL}^q|^2 - \sum_q |\varepsilon_{RR}^q|^2\right) + \left(\sum_q |\bar{\varepsilon}_{LR}^q|^2 - \sum_q |\varepsilon_{RL}^q|^2\right)(1-y)^2}{\left(\sum_q |\bar{\varepsilon}_{LL}^q|^2 + \sum_q |\varepsilon_{RR}^q|^2\right) + \left(\sum_q |\bar{\varepsilon}_{LR}^q|^2 + \sum_q |\varepsilon_{RL}^q|^2\right)(1-y)^2} . \tag{5.21}$$

The amplitudes ε_{kl}^i are identical with those from (5.3) with the obvious replacement $e \to \mu$. The amplitudes $\bar{\varepsilon}_{kl}^i$ are obtained from ε_{kl}^i by observing that both charge and v_μ change sign when going from lepton to antilepton, while a_μ retains its sign:

$$|\bar{\varepsilon}_{LL}^q|^2 = Q_q^2 + Q_q(-v_\mu v_q + a_\mu a_q - v_\mu a_q + a_\mu v_q)\text{Re}\{g(t)\}$$

$$|\bar{\varepsilon}_{LR}^q|^2 = Q_q^2 + Q_q(-v_\mu v_q - a_\mu a_q + v_\mu a_q + a_\mu v_q)\text{Re}\{g(t)\}$$

$$|\varepsilon_{RL}^q|^2 = Q_q^2 - Q_q(+v_\mu v_q - a_\mu a_q + v_\mu a_q - a_\mu v_q)\text{Re}\{g(t)\}$$

$$|\varepsilon_{RR}^q|^2 = Q_q^2 - Q_q(+v_\mu v_q + a_\mu a_q - v_\mu a_q - a_\mu v_q)\text{Re}\{g(t)\} .$$

$$\tag{5.22}$$

With the same approximations as for (5.4), the B asymmetry (see also Berman and Primack, 1974) is given by

$$B(y) = -\frac{G}{\sqrt{2}} \frac{Q^2}{4\pi\alpha} \left(\frac{3}{5}\right) \left(b_1 + b_2\right) \frac{1-(1-y)^2}{1+(1-y)^2} , \tag{5.23}$$

with

$$b_1 = -a_\mu(2a_u - a_d)$$

$$b_2 = +v_\mu(2a_u - a_d) .$$

Since $|a_\mu| \gg |v_\mu|$, as determined in e^+e^- reactions, the B asymmetry is mainly parity-conserving. In deriving (5.23) complete polarisation of the muons has been assumed. For the actual experiment a polarisation P_μ of -0.81 ± 0.04 was measured for the 200 GeV incident μ^+ beam. Therefore the parity-violating part b_2 of the B asymmetry receives a factor $|P_\mu|$, in analogy to the parity-violating asymmetry (5.6):

$$B_{\exp}(y) = -\frac{G}{\sqrt{2}} \frac{Q^2}{4\pi\alpha} \left(\frac{3}{5}\right) \left(-a_\mu + \lambda v_\mu\right) (2a_u - a_d) f(y) , \tag{5.24}$$

with

$$\lambda = |P_\mu| ,$$

$$f(y) = \frac{1-(1-y)^2}{1+(1-y)^2} .$$

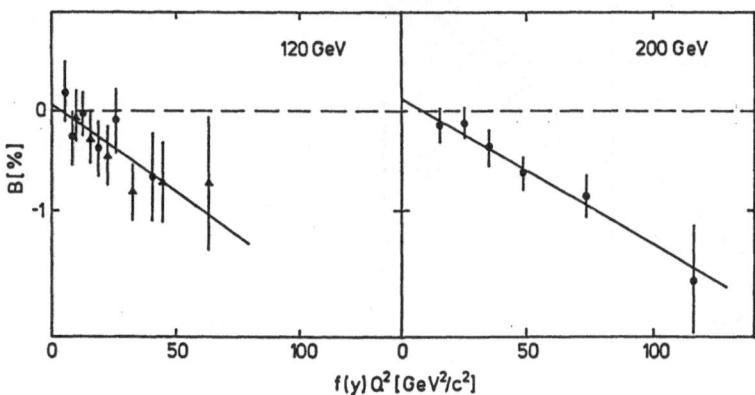

Fig. 5.3. Measurements of the cross section asymmetry $B(y)$ as a function of $Q^2 f(y)$ in μC scattering (Argento et al., 1983). The definition of $f(y)$ is given in the text

The measured asymmetry $B(y)$ receives a major correction from higher order electromagnetic processes (Bardin et al., 1982), which produce a positive charge asymmetry, similar to the one encountered in $e^+ e^-$ reactions. Only here, due to the lower Q^2 values involved, the relative size of the correction with respect to the weak effect is of order 50% at 200 GeV incident muon momentum. The measurements of $B(y)$ after corrections as a function of y are shown in Fig. 5.3. The data have been used to constrain the coefficients b_1 and b_2 from (5.23):

$$\left(-a_\mu + 0.81 v_\mu\right)\left(2a_u - a_d\right) = 2.72 \pm 0.68\,(\text{stat.}) \pm 0.37\,(\text{syst.}) . \qquad (5.25)$$

Relation (5.25) can be confronted in a number of ways with the expectation from the standard theory:

— Assuming the standard couplings for the muon as supported by $e^+ e^-$ scattering yields a measurement for $2a_u - a_d$, expected to be equal to $+3$. The experimental result is 2.91 ± 0.83.

— Assuming the standard couplings for the quarks provides constraints on v_μ:

$$-a_\mu + 0.81 v_\mu = 0.91 \pm 0.23\,(\text{stat.}) \pm 0.12\,(\text{syst.}) .$$

The 1σ limits from this relation are shown in Fig. 5.4 together with the constraints on a_μ from $e^+ e^-$ scattering.

— Using the standard value $a_\mu = -1$ the vector coupling constant of the muon becomes

$$v_\mu = -0.11 \pm 0.28\,(\text{stat.}) \pm 0.15\,(\text{syst.}) .$$

The expectation from the standard theory is $v_\mu = -0.08$ for $\sin^2 \theta_W = 0.23$.

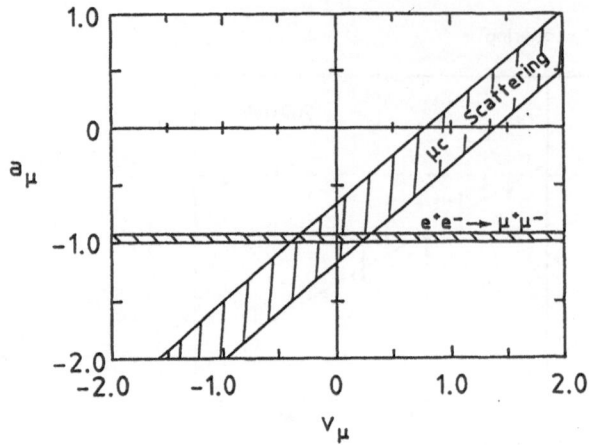

Fig. 5.4. 68% C.L. contours in the $v_\mu - a_\mu$ plane from μC scattering (Argento et al., 1983) and $e^+e^- \to \mu^+\mu^-$ data

— Assuming the standard theory expressions for vector and axial-vector charges the resulting weak mixing angle is given by

$$\sin^2\theta_W = 0.23 \pm 0.07\,(\text{stat.}) \pm 0.04\,(\text{syst.}) \;.$$

The BCDMS Collaboration (Argento et al., 1984) have also measured, for the first time, the interference structure function $x\,G_3(x)$ which was introduced by Derman (1973). In terms of the quark parton model, $x\,G_3(x)$ is given by

$$x\,G_3(x) = x \sum_q a_q\, Q_q\, (q(x) - \bar{q}(x)) \;.$$

The interference structure function is derived from the difference of cross sections with polarised μ_L^+ and μ_R^- on carbon nuclei:

$$\frac{d\sigma(\mu_L^+)}{dQ^2\,dx} - \frac{d\sigma(\mu_R^-)}{dQ^2\,dx} = \frac{G}{\sqrt{2}}\frac{\alpha}{Q^2\,x}\,(a_\mu - \lambda v_\mu)\,(1 - (1-y)^2)\,x\,G_3 \;.$$

Normalising to the electromagnetic structure function $F_2(x)$ which, in the simple quark model, is given by

$$F_2(x) = x \sum_q Q_q^2\,(q(x) - \bar{q}(x)) \;,$$

one finds, for large x where sea quark contributions can safely be neglected, the following relation:

$$\frac{x\,G_3(x)}{F_2(x)} = \frac{a_u\,Q_u + a_d\,Q_d}{Q_u^2 + Q_d^2} \;. \tag{5.26}$$

In the standard theory this ratio is predicted to be 9/5. Comparing the two structure functions $F_2(x)$ and $x\,G_3(x)$, the BCDMS Collaboration has determined the above ratio of structure functions as 1.87 ± 0.25 (stat.) ± 0.42 (syst.). This measurement can be used to exclude regions in the $u_R - d_R$ plane shown in Fig. 4.12, remembering the relation between the axial-vector charge and the chiral couplings (2.36)

$$a_u = 2(u_L - u_R) \,, \quad a_d = 2(d_L - d_R) \,.$$

Taking the measurements for u_L and d_L from neutrino nucleon scattering (see Sect. 4.3), the μC data lead to the boundaries indicated in Fig. 4.12 and thus support the preferred solution from the neutrino data, in splendid agreement with the standard therory.

6. The Weak Bosons

As outlined in Chap. 2, the standard theory, once the Weinberg angle $\sin^2 \theta_W$ is known, makes a definite prediction for the masses of $W^\pm$ and Z^0 (see (2.42)). Using $\sin^2 \theta_W = 0.23$, masses of about 80 and 90 GeV/c^2 are predicted for $W^\pm$ and Z^0, respectively. At the time when these rough estimates were available there was no accelerator at hand to provide enough energy to produce these particles, if they existed. The CERN SPS, e.g., planned to operate at a maximum of 400 GeV, would only allow for a centre of mass energy of $\sim$ 28 GeV on a stationary proton target. In a technologically extremely demanding effort the CERN SPS has consequently been converted into a colliding beam storage ring for protons and antiprotons (Van der Meer, 1972 and Rubbia et al., 1976, the Staff of the CERN $p\bar{p}$ Project, 1981) providing a total centre of mass energy up to 630 GeV (315 GeV per beam) with good luminosity. In the quark-parton picture, this energy is shared in roughly equal parts by valence quarks and gluons, so that the average total centre of mass energy of a quark-antiquark collision is of order M_Z. The reactions in which one hoped to discover the intermediate bosons were (taking the W^- for definiteness)

$$\bar{p}p \to W^- X$$
$$\hookrightarrow e^- \bar{\nu}_e, \ \mu^- \bar{\nu}_\mu \ , \tag{6.1}$$

and

$$\bar{p}p \to Z^0 X$$
$$\hookrightarrow e^+ e^-, \ \mu^+ \mu^- \ ,$$

which correspond to the elementary processes (see Fig. 6.1)

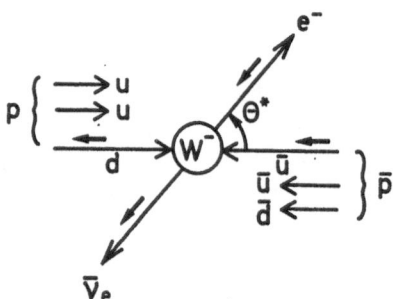

Fig. 6.1. Elementary scattering partners in $\bar{p}p$ reactions and definition of the emission angle for the electron from W^- decays

$$\bar{u}d \rightarrow W^- \rightarrow e^- \bar{\nu}_e, \; \mu^- \bar{\nu}_\mu \, ,$$

$$\bar{q}q \rightarrow Z^0 \rightarrow e^+ e^-, \; \mu^+ \mu^- \, . \tag{6.2}$$

Finding these bosons in a series of extraordinary experiments was certainly the major step towards the present, very satisfactory understanding of the forces between the elementary constituents of matter.

6.1 The UA Experiments

Two experiments have searched for the intermediate bosons, the UA1 collaboration (Arnison et al., 1983a) and the UA2 collaboration (Banner et al., 1983), who have pursued quite different detector concepts.

The UA1 detector is a hermetic 4π detector, covering the polar angle range down to 0.2 degrees with respect to the incident beams. The momenta of charged particles are analyzed in a 0.7 Tesla dipole field, transverse to the beam, equipped with a pictorial drift chamber for tracking. The electromagnetic calorimeter (27 radiation lengths of lead-scintillator) surrounding the central drift chamber assembly is longitudinally subdivided into 4 samplings and followed by a hadron calorimeter (iron-scintillator). Muons are detected in a range telescope following the hadron calorimeter. The absolute energy calibration of the calorimeter, the essential tool to measure the masses of the intermediate bosons, is known within 3% (Arnison et al., 1983a–c, 1985).

UA2, on the other hand, have emphasised calorimetric measurements with their detector. The central part of the detector covering the polar angle interval $40° < \theta < 140°$ with respect to the incident beams consists of a tracking detector and preshower counters, followed by a non-magnetic calorimeter, optimised for precise angular and energy measurements of photons and electrons. The calorimeter is segmented longitudinally into an electromagnetic (17 radiation lengths lead-scintillator) and a hadronic (2 absorption lengths iron-scintillator) part. The forward regions (down to 20°) are equipped with toroidal magnet spectrometers (0.38 Tesla m) and electromagnetic calorimeters. The absolute energy calibration for UA2 is known within 1.6% (Bagnaia et al., 1984; Appel et al., 1985).

For the search of $W^\pm$ and Z^0 decays in the debris of violent collisions of protons with antiprotons, good identification of electrons is essential. The UA1 calorimeter provides four longitudinal samplings for electromagnetic showers so that a comparison can be made with the expected shower profile for electrons. Because of the limited lateral granularity of their calorimeter UA1 have to require isolation in space for an electron. Typically less than 3 GeV transverse energy (or momentum) from other charged particles are required in a cone of 40° opening angle around the electron candidate. UA2, on the other hand, have only one longitudinal sampling in the electromagnetic calorimeter, but good lateral granularity. One therefore compares with the expected *lateral* shower profile for electrons, with a less stringent spatial isolation requirement. In addition the identification of electrons is aided by the requirement of small energy leakage into the hadronic calorimeter.

In principle, the existence of non-interacting particles such as neutrinos can be inferred in a given event by substantial "missing" energy, deduced by summing up the energy of all detected particles and establishing a deficit with respect to the total centre of mass energy provided by the beams. However, in a typical $\bar{p}p$ interaction a large fraction of the produced particles is emitted along the beam direction and thus remains undetected. One therefore resorts to the measurement of the missing transverse momentum $\not{p}_T$ by requiring conservation of the total transverse momentum. The UA1 calorimeters cover the scattering angle interval down to 0.2 degrees as opposed to UA2 (20 degrees) and consequently can perform a better measurement on $\not{p}_T$ or $\not{E}_T$.

Data were taken by both experiments in two running periods. During 1982-1983 each experiment collected data at $\sqrt{s} = 546$ GeV, while in 1984 the machine was set to $\sqrt{s} = 630$ GeV. The total integrated luminosities at the two energy settings, on which the results in this chapter are based, are 136 and 630 nb^{-1} for UA1, and 142 and 768 nb^{-1} for UA2.

Both experiments stopped running in 1986 and have since then been subjected to a major upgrading programme (UA1: Dowell, 1986; UA2: Booth, 1986). With improved antiproton currents provided by the new antiproton collector ACOL (Jones et al., 1983) the upgraded detectors are scheduled to resume data taking in late 1988.

6.2   Production and Decay Properties

The production mechanisms and experimental signatures for $W^{\pm}$ and Z^0 production are evident from (6.1, 2): Valence quarks and antiquarks produce a $W^{\pm}$ or Z^0 with generally small transverse momentum p_T, a known fraction of which subsequently decay into lepton pairs with large p_T. The observable rate is thus proportional to the product of the production cross section, calculable in a QCD model for the proton (antiproton) structure functions with standard couplings of the weak bosons to the quarks, and the branching ratios into leptonic final states (see (2.44, 45)). Such calculations have been carried out by Altarelli et al. (1984, 1985), indicating that the rate for $W^{\pm}$ production with subsequent decay into $e\nu$ is about 10 times larger than the rate for Z^0 production with subsequent decay into an e^+e^- pair.

Table 6.1 shows the measurements from both UA experiments together with the Altarelli et al. predictions. Reasonable agreement between theory and experiment is observed. It should be noted that the theoretical uncertainties in the calculations mainly come from input outside of the standard theory, such as proton structure functions and higher order QCD corrections (the famous "K" factor). $W^{\pm}$ production is signalled by electrons (positrons) with large p_T ("Jacobian peak" at $\sim \frac{1}{2} M_W$) and a large p_T neutrino, to be identified by substantial

Table 6.1. Measurements for $\sigma \cdot$ BR for $W^{\pm}$ and Z^0 production in nb from the UA experiments. The quoted errors are statistical and systematic Also shown are the expectations given by Altarelli et al.(1984, 1985). The theoretical errors estimate the uncertainties, mainly due to the incomplete knowledge of the structure proton functions and higher order QCD corrections

	$\sqrt{s}$	UA1	UA2	Theory
W	546	$0.55 \pm 0.08 \pm 0.09$	$0.61 \pm 0.10 \pm 0.07$	$0.37^{+0.12}_{-0.05}$
	630	$0.63 \pm 0.05 \pm 0.10$	$0.57 \pm 0.04 \pm 0.07$	$0.47^{+0.14}_{-0.08}$
Z	546	$0.042^{+0.033}_{-0.020} \pm 0.006$	$0.116 \pm 0.039 \pm 0.011$	$0.042^{+0.013}_{-0.007}$
	630	$0.074 \pm 0.014 \pm 0.011$	$0.073 \pm 0.014 \pm 0.007$	$0.051^{+0.016}_{-0.010}$

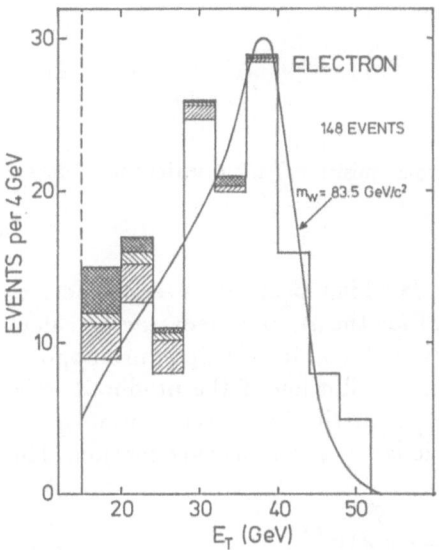

Fig. 6.2. Transverse energy distribution for electrons (positrons) from W decays. The shaded areas indicate various background contributions indicated in the text

missing transverse energy E_T. The hadronic system X in (6.1) resulting from the fragmentation of the non-interacting ("spectator") quarks is emitted at small p_T. Searches for $\bar{p}p$ reactions containing a well-identified, isolated, large p_T electron (positron) and a large amount of missing transverse energy gave first evidence for W production (Arnison et al., 1983a). Shortly afterwards UA2 (Banner et al., 1983) have confirmed the discovery. Requiring in essence the high E_T (typically 25 GeV) of the electron to be matched by the missing E_T results in rather clean W samples, with background from jet fluctuations and more complex W decays of order $< 10\%$. Figure 6.2 shows the distribution of the transverse energy for electrons (positrons) from the W sample of the UA1 experiment (Arnison et al., 1985).

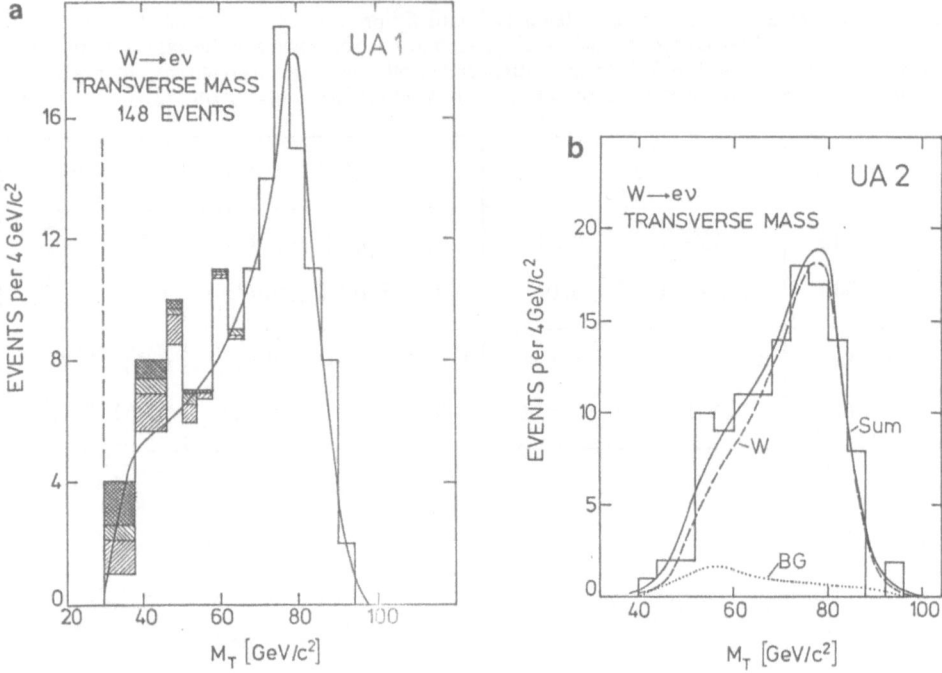

Fig. 6.3. Transverse mass distribution from electrons (positrons) and neutrinos in W decays from UA1 and UA2 experiments

The E_T distribution clearly exhibits a Jacobian peak at an energy corresponding to roughly $\frac{1}{2} M_W$. A cleaner signal for the W can be extracted calculating the transverse mass m_T of the W boson. This, in principle, presupposes the knowledge of the transverse momentum distribution of the produced W's. If, however, the missing transverse energy $\not{E}_T$ is attributed to the ν from the W decay, m_T can be determined without knowledge of the transverse motion. The transverse mass m_T is defined as

$$m_T = (2 p_T^e \, p_T^\nu \, (1 - \cos \phi))^{1/2}$$

where ϕ is the angle between the transverse momentum vectors of e and ν.

Figure 6.3 shows examples of the transverse mass distributions from UA1 (Albajar et al., 1987) and UA2 (Ansari et al., 1987). Note that the distributions can be made more symmetric by raising the minimum E_T requirement for the electrons. Mass and width of the W can be determined by comparing model calculations of W production and decay to the data. Both UA experiments performed Monte Carlo calculations taking into account the expected W motion in the laboratory frame and the known detector efficiencies as well as backgrounds from the above-mentioned sources. The distributions of the transverse mass after simulation were then adjusted to the data in fits leaving the mass M_W and the width Γ_W as free parameters. Table 6.2 summarises the results of these fits.

Table 6.2. Measurements of the mass and limits on the width for the $W^\pm$ boson. The first error on the W mass given in the table is statistical, the second systematic

Experiment	Channel	M [GeV/c^2]	Γ [GeV/c^2]
UA1	$W \to e\,\nu$	$82.7 \pm 1.0 \pm 2.7$	< 6.5
	$W \to \mu\,\nu$	$81.8^{+6.0}_{-5.3} \pm 2.6$	(90% C.L.)
	$W \to \tau\,\nu$	$89 \pm 3 \pm 6$	
UA2	$W \to e\,\nu$	$80.2 \pm 0.6 \pm 1.4$	< 7.0
			(90% C.L.)

UA1, due to their excellent resolution in the missing transverse energy and their ability to measure charged particle momenta, have found the decay of W bosons into a muon and a neutrino (see Table 6.2). They have also isolated events consistent with the decay

$$W \to \tau\,\bar{\nu}_\tau,\ \tau \to \nu_\tau + \text{hadrons} .$$

The events are characterised by a single narrow jet, approximately back-to-back with significant missing energy. The narrow jets have low charged multiplicity, low invariant mass, and substantial energy deposition, suggesting the observation of

$$\tau \to \nu_\tau + \pi + n\,\pi^0 ,$$

$$\tau \to \nu_\tau + 3\pi + n\,\pi^0 .$$

Calculations have shown that the selection criteria employed for the τ search efficiently isolate the τ-decay sample from other potential mono-jet signatures. Taking into account the respective selection and reconstruction efficiencies for the three leptonic decay channels, UA1 find excellent agreement with universality. Defining

$$\frac{\Gamma(W \to l_i\nu)}{\Gamma(W \to l_j\nu)} = \left(\frac{g_i}{g_j}\right)^2 , \tag{6.3}$$

the following results are obtained:

$$\frac{g_\mu}{g_e} = 1.00 \pm 0.07 \pm 0.04 , \quad \frac{g_\tau}{g_e} = 1.01 \pm 0.10 \pm 0.06 . \tag{6.4}$$

One can further determine the branching ratio (BR) of the W into the individual leptonic channels by assuming the production cross section as given by the model calculations (see Table 6.1). For the decay $W \to e\nu$ the result is

$$BR(W \to e\nu) = 0.10 \pm 0.014 \,^{+0.02}_{-0.03} \,, \tag{6.5}$$

with rather large systematic errors arising from the uncertainties in the QCD model.

Concerning the mass value obtained in the $e\nu$ decay mode, both experiments agree well. Taking the weighted mean from both experiments (adding statistical and systematic errors in quadrature), the mass of the W is determined to be

$$M_W = 80.7 \pm 0.5 \,(\text{stat.}) \pm 1.2 \,(\text{syst.}) \,\, [\text{GeV}/c^2] \,. \tag{6.6}$$

This value is in perfect agreement with the lowest order expectation of the standard theory (see (2.43)), where the value for $\sin^2 \theta_W$ has been taken equal to 0.23 as suggested by the measurements from deep inelastic neutrino nucleon scattering (see Chap. 4). More discussion on this point will follow.

Recently, the CDF detector at the FNAL collider, running since 1987, has reported first results on the W production in $\bar{p}p$ collisions at 1.8 TeV (Errede, 1988). CDF obtain 22 W candidates, virtually free of background, and find a mass of $M_W = 80.0 \pm 3.3 \,(\text{stat.}) \pm 1.2 \,(\text{syst.}) \,\, [\text{GeV}/c^2]$, in good agreement with the UA experiments. CDF have also determined the production cross section for W decaying into $e\nu$ and find $\sigma_W \cdot BR(W \to e\nu) = 2.6 \pm 0.6 \pm 0.5$ nb, also at the upper end of the Altarelli et al. calculation (see Table 6.1).

Apart from the prediction of the mass of the W, which has so impressively been verified, yet another piece of information, namely the spin $J = 1$ assignment of the W, clearly demonstrates the correct interpretation of the signal. As the W's are produced at the collider energies predominantly in valence quark-antiquark annihilation one expects a pronounced angular asymmetry of the electron (positron) emission in the W rest frame with respect to the beam direction. The $W^\pm$ couples to left-handed particles (right-handed antiparticles) only, it is therefore longitudinally polarised with its spin aligned with the antiproton direction of flight (see Fig. 6.1). Due to conservation of angular momentum, the decay angular distribution of the electron (positron) is proportional to $(1 + \cos \theta^*)^2$, where θ^* is the angle between the electron (positron) and proton (antiproton) directions in the $W^-(W^+)$ rest system. Both UA experiments (Arnison et al., 1985; Appel et al., 1985) have investigated the decay angular distributions. Since only the decay electron and the transverse momentum of the decay neutrino are measured the angle θ^* cannot be calculated without further assumptions. Constraining the invariant mass of electron and neutrino to the mass of the W, two solutions for the longitudinal momentum of the neutrino are obtained, out of which one usually is unphysical. In about 3/4 of the W candidates a unique physical solution is obtained. Figure 6.4 shows the decay angular distributions of W candidate subsamples from both UA experiments, where the laboratory momentum of the W could be unambigously reconstructed allowing a Lorentz

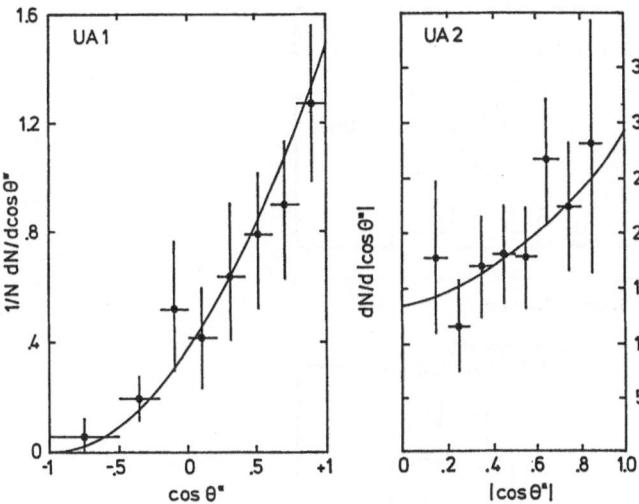

Fig. 6.4. Angular distribution on electrons from W decays in the W rest frame from UA1 and UA2 experiments

transformation of the observed $e^\pm$ into the W rest system. Since UA2 do not have magnetic analysis their distribution is symmetric around $\cos\theta^* = 0$, corresponding to $(1 + \cos^2\theta^*)$ for completely polarised W's. Both angular distributions are in agreement with the expectation of a completely polarised $J = 1$ particle, decaying via the $V - A$ interaction.

More quantitatively, it has been shown by Jacob (1958) that for a particle of arbitrary spin J one expects for a two body production and decay

$$\langle\cos\theta^*\rangle = \frac{\langle\mu\rangle\,\langle\lambda\rangle}{J\,(J+1)}\;,$$

where $\langle\mu\rangle$ and $\langle\lambda\rangle$ are the total initial and final state helicities, respectively. For W production and decay which both proceed through the $V - A$ interaction one has

$$\langle\lambda\rangle = -1\;,$$

for the $u\bar{d}$ or $d\bar{u}$ initial state and

$$\langle\mu\rangle = -1\;,$$

for the $e^+\nu_e$ or $e^-\bar{\nu}_e$ the final state. The experimental result from UA1 is

$$\langle\cos\theta^*\rangle = 0.43 \pm 0.07$$

which, compared to 0.5 expected for $J = 1$, supports the assignment of $J = 1$ for the W gauge bosons.

Turning now to the neutral partner of the $W^\pm$ bosons, the signature for Z^0 production is clean and simple from an experimental point of view, even in $p\bar{p}$

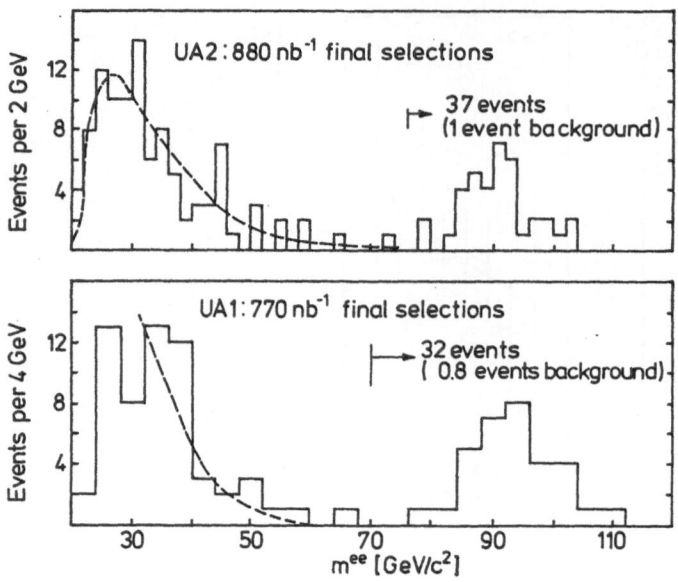

Fig. 6.5. Invariant mass of electron pairs for UA1 and UA2 experiments

collisions: One looks for two charged particles, identified as electrons, forming a high invariant mass, typically in excess of 50 GeV/c^2. Monte-Carlo calculations show that the stringent selection requirements for electrons and positrons together with the high invariant mass requirement lead to a virtually background-free Z^0 sample. UA1 (Arnison et al., 1985), and UA2 (Appel et al., 1985) have found 16 Z^0's each in their data excluding the 1985 running. The full data samples contain almost 40 events each. The actual number of candidates depends on the specific particle identification criteria. As an example of such selections, the invariant mass for two-electron candidates above a mass of 20 GeV are shown in Fig. 6.5. Both experiments see unambiguous accumulations of events at the expected Z^0 mass, almost free of background.

In a subsample of their data UA1 have searched for the decay $Z^0 \rightarrow \mu^+\mu^-$. In an earlier publication (Arnison et al., 1983b) they observe one event, in agreement with the expectation of universal coupling. The full data sample contains 19 events. The observed rate of $\mu^+\mu^-$ events is in very good agreement with universality. Forming a ratio similar to (6.3) UA1 obtain

$$\frac{g_{\mu\mu}}{g_{ee}} = 1.02 \pm 0.15 \pm 0.04 \ .$$

The results from both experiments, concerning $Z^0 \rightarrow l^+l^-$, are summarised in Table 6.3. The mass of the Z^0 is obtained from fitting a relativistic Breit-Wigner shape to the invariant mass distribution. The width Γ_Z can be directly measured from the width of the experimental mass peak. The determination, however, depends critically on the precise knowledge of the measurement accuracy. Therefore

Table 6.3. Measurements on the Z mass. The first error on the Z mass given in the table is statistical, the second systematic

Experiment	Channel	M [GeV/c^2]	Γ [GeV/c^2]
UA1	$Z \to e^+e^-$	$93.1 \pm 1.0 \pm 3.1$	< 8.3
	$Z \to \mu^+\mu^-$	$90.7^{+5.2}_{-4.8} \pm 3.2$	(90% C.L.)
UA2	$Z \to e^+e^-$	$91.5 \pm 1.2 \pm 1.7$	< 7.1
			(90% C.L.)

only limits are given in the table. As for the W's, UA2 quote smaller systematic errors for the mass determination of the Z^0, due to their better control of the absolute energy calibration which is routinely verified in seperate beam tests. The agreement between the two experiments concerning the mass values of the Z^0 is good. Taking the weighted average one obtains for the Z^0 mass

$$M_Z = 92.0 \pm 0.8 \text{ (stat.)} \pm 1.5 \text{ (syst.)} \text{ [GeV/}c^2\text{]} . \qquad (6.7)$$

Also the mass of the Z^0 is in remarkable agreement with the prediction from (2.43). No results on the Z^0 have been reported from FNAL as of yet, which is not too surprising: Assuming the ratio 10 for $(\sigma \cdot \text{BR})_W/(\sigma \cdot \text{BR})_Z$ CDF should have seen only 2 Z^0 events in their statistics of 25.3 nb^{-1}.

In contrast to W decays, electrons from $Z^0 \to e^+e^-$ are not expected to show a strong asymmetry in the decay angular distribution. For $\sin^2 \theta_W = 1/4$, the asymmetry is predicted to vanish identically, as right- and left-handed couplings are equal in magnitude (see Table 2.1). The decay angular distribution for the UA1 Z^0 events is shown in Fig. 6.6, together with predictions for $\sin^2 \theta_W = 0.25$ (symmetric) and $\sin^2 \theta_W = 0.214$ (small positive asymmetry). Clearly, the lim-

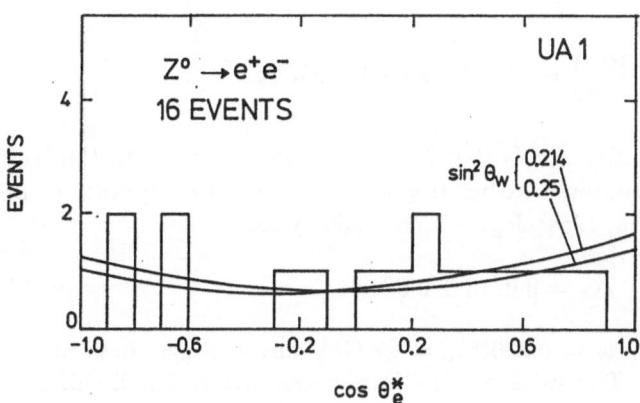

Fig. 6.6. Decay angle of the electron from Z^0 candidates with respect to the proton beam axis

ited statistics does not allow any strong conclusion. The distribution, however, is compatible with the almost symmetric case expected in the standard theory for $\sin^2 \theta_W \approx 0.23$.

Owing to their high centre of mass energy, the collider experiments have the best potential to detect another elusive particle predicted by the standard theory, the t quark. For t quark masses smaller than the W very clean signatures are expected for semileptonic decays. All searches for the top, however, have been unsuccessful so far (see, e.g., Jenni (1987) and references contained therein).

6.3 Analysis of the Boson Masses

Sticking to the well-founded hypothesis that the observed heavy particles are indeed the desired weak quanta $W^\pm$ and Z^0 we now proceed to a more detailed discussion of the UA results within the standard theory. The most important parameter to be extracted from the boson masses is the Weinberg angle, the definition of which depends on the renormalisation scheme chosen. Using the on-shell renormalisation scheme the Weinberg angle is defined by

$$\sin^2 \theta_W = 1 - \frac{M_W^2}{M_Z^2} .$$ (6.8)

In that scheme the determination of $\sin^2 \theta_W$ is very direct, independent of other experimental input, and completely free of theoretical uncertainties (compare, e.g., the $\sin^2 \theta_W$ determinations in the preceeding chapters). From an experimental point of view, the definition (6.8) is also advantageous since the uncertainties in the absolute energy calibration of the calorimeters cancel to first order.

A more precise measurement for $\sin^2 \theta_W$, however, can be obtained using the relations (2.40, 41):

$$\sin^2 \theta_W = \frac{A^2}{M_W^2 (1 - \Delta r)}$$ (6.9)

with

$$A = \left(\frac{\pi \alpha}{G\sqrt{2}} \right) = 37.2810 \pm 0.0003 \text{ [GeV]} .$$

In this expression the quantity Δr is the $O(\alpha)$ one-loop correction to the tree level formula for M_W encountered earlier (see (2.40)). Recent computations of Δr (see Marciano and Sirlin, 1984; Jegerlehner, 1986) yield

$$\Delta r = 0.0713 \pm 0.0013 ,$$ (6.10)

where the parameters $\sin^2 \theta_W = 0.230, m_t = 45$ GeV and a Higgs mass $m_\phi = 100$ GeV have been used. The value for Δr is not sensitive to small shifts in

Table 6.4. Measurements for $\sin^2 \theta_W$, ϱ, and the radiative correction term Δr from the $\bar{p}p$ collider. The equation numbers according to which the values are determined are given in the first column

$\sin^2 \theta_W$	UA1	UA2
(6.8)	0.211 ± 0.025	$0.232 \pm 0.025 \pm 0.010$
(6.9)	$0.218 \pm 0.005 \pm 0.014$	$0.232 \pm 0.003 \pm 0.008$
ϱ	UA1	UA2
(6.11)	$1.009 \pm 0.028 \pm 0.020$	$1.001 \pm 0.028 \pm 0.006$
Δr	UA1	UA2
(6.12)	$0.038 \pm 0.100 \pm 0.067$	$0.068 \pm 0.087 \pm 0.030$
(6.13)	$0.125 \pm 0.021 \pm 0.057$	$0.068 \pm 0.022 \pm 0.032$

$\sin^2 \theta_W$, m_ϕ or m_t. The theoretical error, however, does not apply for substantial shifts, e.g. $m_t \sim O\left(200 \text{ GeV}/c^2\right)$.

Relations (6.8, 9) can be used to derive the value of $\sin^2 \theta_W$ from the boson masses. A straightforward calculation shows that, although the absolute energy scale errors largely cancel in relation (6.8), the precision of the result is far inferior to using M_W alone. In fact, one obtains

$$\frac{\Delta \sin^2 \theta_W}{\sin^2 \theta_W} = \frac{2}{M_W} \left(\frac{M_Z^2}{M_W^2} - 1\right)^{-1} \sqrt{(\Delta M_W)^2 + \frac{M_W^2}{M_Z^2} (\Delta M_Z)^2}$$

$$\approx \frac{2\Delta M_W}{M_W} \frac{\sqrt{2}}{\sin^2 \theta_W}$$

from relation (6.8), and

$$\frac{\Delta \sin^2 \theta_W}{\sin^2 \theta_W} = \frac{2\Delta M_W}{M_W}$$

from relation (6.9), so that the error on $\sin^2 \theta_W$ from (6.8) is expected to be about a factor of 6 larger compared to the one from (6.9). The values for $\sin^2 \theta_W$ obtained by UA1 and UA2 using either method are given in Table 6.4. All measurements are in good mutual agreement and also compare favourably to the precise determination from neutrino nucleon scattering, where $\sin^2 \theta_W = 0.233 \pm 0.006$ was measured (see Sect. 4.3).

Also given in the table are the measurements for the ϱ parameter, which is unity in the minimal model with only one neutral Higgs boson (see Sect. 2.3). ϱ is defined by the relation

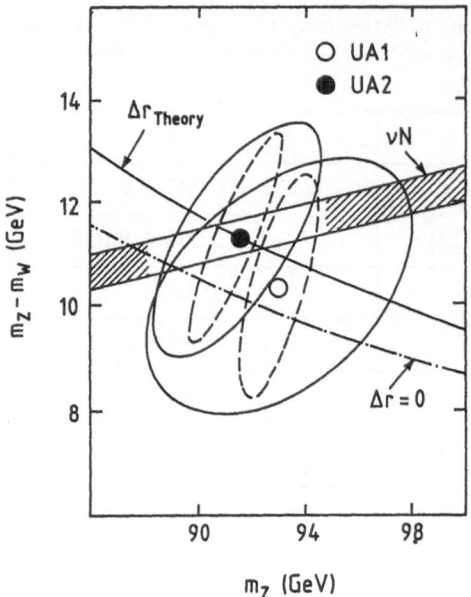

Fig. 6.7. Relation between $M_Z - M_W$ and M_Z. The standard theory prediction, including the one-loop radiative correction, is given by the solid curve. The Born level is labelled $\Delta r = 0$. 68% confidence level contours for the data are shown. The dashed ellipses correspond to the statistical errors only, the solid ellipses correspond to the statistical and systematic error added in quadrature. The band shows the region allowed by the neutrino nucleon measurement of $\sin^2 \theta_W$

$$\varrho = \frac{M_W^2}{M_Z^2 \cos^2 \theta_W} \, , \tag{6.11}$$

where $\cos^2 \theta_W$ is derived from relation (6.9). Both experiments are in good agreement with $\varrho = 1$ and with the equivalent determinations from the low energy experiments.

The relation (6.9) can also be turned around to determine the radiative correction Δr itself from experiment. In doing so one is able, in principle, to test the standard theory at the one-loop level by a measurement of Δr:

$$\Delta r = 1 - \frac{A^2}{M_W^2 (1 - M_W^2 / M_Z^2)} \, . \tag{6.12}$$

Using the boson masses in (6.12) leads to a determination of Δr which is clearly compatible with zero, as can be seen from Table 6.4. The reason is that the low precision determination of the Weinberg angle (6.8) has been used. Alternatively, one can define

$$\Delta \tilde{r} = 1 - \frac{A^2}{M_W^2 \sin^2 \theta_W (\nu N)} \, , \tag{6.13}$$

and insert the precise value for $\sin^2 \theta_W$ from ν nucleon scattering. In this case a significant result for $\Delta \tilde{r}$ is obtained, averaging over both experiments:

$$\Delta \tilde{r}_{\mathrm{Exp}} = 0.084 \pm 0.033 \, . \tag{6.14}$$

This point can further be illustrated by studying the mass difference $\Delta_M = M_Z - M_W$, as suggested by Marciano and Sirlin (1984). The influence of the radiative corrections on Δ_M as a function of M_Z is shown in Fig. 6.7. where the curves are calculated for the tree level and the one-loop level, respectively. Also drawn are the 68% C.L. contours for the two UA experiments as well as the range allowed by the neutrino nucleon experiments. It is clear from this figure that the existence of radiative corrections cannot be demonstrated solely on the basis of the $\bar{p}p$ experiments. Only when combining the $\bar{p}p$ experiments with neutrino nucleon scattering (see (6.14)) may one be tempted to conclude that electroweak radiative corrections have indeed been observed.

The discovery of the intermediate vector bosons and the measurement of their masses have impressively verified the most important features of the standard theory in its minimal form with a Higgs doublet ($\varrho = 1$). Given the complete success of the standard theory in predicting the observed phenomena, early results on the Z^0 have uncovered peculiar decays, such as mono-jets, Z^0 decays into $e^+e^-\gamma$ and others. With more statistics, however, these strange effects have not appeared again or have been reinterpreted as originating from conventional sources, The standard theory still stands, with no serious competitor in sight.

7. The Parameters of the Standard Theory

The most important single experimental support of the GSW theory of electroweak interactions certainly is the discovery of the intermediate heavy bosons $W^{\pm}$ and Z^0 (see Chap. 6). In addition, there is a wealth of data on weak neutral current interactions of leptons and quarks, spanning a Q^2 range of about 10(!) orders of magnitude, from atomic spectroscopy (few MeV^2 in Cs) to the production of Z^0 from $q\bar{q}$ pairs ($\sim 10^4\,\text{GeV}^2$). All these data are impressively well described by the GSW theory.

There is information on the couplings of the neutral current to the charged leptons of all three generations. Also the u and d quark couplings have been well measured in lepton-nucleon scattering experiments, while little is known about the couplings for quarks of the second and third generation.

The experimental situation concerning the fermion couplings to the weak neutral current is shown in Table 7.1, where the most precise determination has been selected for each case. Note that the chiral couplings are related to the vector and axial-vector couplings by means of relation (2.36). The couplings of the electron are taken from neutrino electron scattering (see Sect. 4.2), while the couplings for the μ and τ are all from e^+e^- scattering (Sect. 3.5), with the exception of the vector coupling of the μ, which is derived from μC scattering (Sect. 5.3). The couplings for the quarks of the first generation come exclusively from neutrino nucleon scattering (Sect. 4.3), and the axial-vector couplings for the quarks of the second and third generation are again from e^+e^- reactions.

In the standard theory all these couplings can be calculated from the knowledge of a single parameter, the weak mixing angle $\sin^2\theta_{\text{W}}$. Another parameter, ϱ, fixes the overall strength of the weak neutral current with respect to the weak charged current. As can be judged from Table 7.1, the measurements agree in great detail with the predictions, assuming $\sin^2\theta_{\text{W}} = 0.23$. In some cases, however, the uncertainties of the measurements are quite substantial.

Both the Weinberg angle and ϱ have to be determined from experiment and an impressive number of measurements, utilising almost all possible fermion scattering processes, have been carried out. Another way of checking the consistency of couplings is therefore to compare the values for $\sin^2\theta_{\text{W}}$ derived from experiments probing different weak neutral current interactions. Before doing so, however, we will study in more detail the assumptions on which the standard theory is based, i.e. universality and factorisation of the neutral current couplings.

Table 7.1. Summary of weak coupling parameters for leptons and quarks. The GSW predictions are calculated with $\sin^2\theta_W = 0.23$

Gen.	1	2	3	
	e	μ	τ	GSW
v	-0.076 ± 0.094	-0.11 ± 0.32	-0.89 ± 1.13	-0.08
a	-0.990 ± 0.052	-1.043 ± 0.056	-0.944 ± 0.093	-1.00
	u	c	t	
L	0.339 ± 0.017	$*$		0.35
R	0.172 ± 0.014	$*$		-0.15
v	0.334 ± 0.062			0.39
a	1.022 ± 0.062	1.12 ± 0.22		1.00
	d	s	b	
L	-0.429 ± 0.014	$**$		-0.42
R	-0.011 ± 0.069	$**$		0.08
v	-0.880 ± 0.166			-0.69
a	-0.836 ± 0.166		-1.0 ± 0.28	-1.00

$* \ (c_L^2 + c_R^2)/(u_L^2 + u_R^2) = 2.1 \pm 1.0$ $\qquad ** \ (s_L^2 + s_R^2)/(d_L^2 + d_R^2) = 1.39 \pm 0.43$

7.1 Universality

The standard theory makes the fundamental assumption that the weak neutral current couples to the fermions only according to their weak isospin and hypercharge assignments, and does not, in particular, distinguish between leptons and quarks nor between generations. This universality of the weak coupling is checked in a model-independent way comparing, e.g., the vector and axial-vector charges for the three generations. Note that the v, a couplings in the first generation are exclusively derived from ν-induced reactions, while the couplings for the other two generations are exclusively derived from e^+e^- annihilation. In order to quantify the level at which universality is verified, let us rather determine at what level universality might be violated, allowing for a one standard deviation in the coupling constant to be tested.

Concerning the axial-vector coupling constants of the leptons, the only well-tested case is $\mu - e$ universality. But even here violations at the level of 10 % are possible when the value of a_μ is allowed to shift by 1σ. The universality test is weaker for the τ, where a violation of up to 15 % is allowed. Finally, $\mu - \tau$

universality may be violated by as much as 22 %. Universality of the lepton vector charges is only very poorly verified. Almost arbitrarily large violations of universality are possible, accepting the 1σ criterium.

In the quark sector a similar situation is found. Although all couplings are consistent with universality, it is not a well tested property. Comparing the axial-vector charges for the u and c quark, violations up to 30 % are admitted, in the case of the d and b quark even up to 50 %. Universality of the vector charges is not tested at all. Furthermore, combinations of v and a couplings for c and u quarks can be studied in NC ψ production (Abramovicz et al., 1981) and lead to the following test

$$(v_c^2 + a_c^2)/(v_u^2 + a_u^2) = 2.1 \pm 1.0 \,,$$

for which a ratio equal to 1 is expected, the experimental error allowing a large violation of universality. Similarly, studying the y distribution in NC current neutrino nucleon reactions, which is influenced by contributions from the sea quarks, an estimate for the strange quark couplings has been derived (Jonker et al., 1981a), assuming standard couplings for u and d quarks:

$$(v_s^2 + a_s^2)/(v_d^2 + a_d^2) = 1.39 \pm 0.43 \,.$$

Also here a ratio equal to 1 is expected, with, obviously, enough room for large violations of universality.

According to Table 7.1, circumstantial evidence for a universal coupling of the weak neutral current to the three generations of quarks and leptons is overwhelming. However, as the simple analysis shows, universality is by no means a well-tested property.

7.2 Factorisation

Another basic assumption of the standard theory is related to the way the weak neutral current is mediated. If there is only one Z^0 boson exchanged, the overall coupling factorises into two components associated with the currents which are coupled by the Z^0 (Hung and Sakurai, 1977). The factorisation property thus relates reactions having one current in common. Three independent factorisation relations exist and may be tested. Ratios rather than the coupling constants themselves are chosen for the tests in order to dispose of the dependences on the parameter ϱ. Assuming lepton universality for e^+e^- interactions, the ratios v^2/a^2 as obtained from e^+e^- and νe scattering can be compared:

$$\left(\frac{v^2}{a^2}\right)_{e^+e^-} = \left(\frac{v^2}{a^2}\right)_{\nu e} \tag{7.1}$$
$$< 0.14 \qquad\quad < 0.014 \qquad (95\% \,\text{C.L.}) \,.$$

As is evident, e^+e^- scattering where the vector charges are only poorly constrained is not precise enough for a serious test of factorisation. A second relation involves the ratio (see (4.30))

$$\frac{\gamma}{\alpha} = \frac{v_u + v_d}{v_u - v_d}$$

of quark couplings as extrated from νq and eD scattering + atomic parity violation in Cs (APV). By combining the vector part of (5.7) from the ed experiment and (5.19) from APV one obtains rough agreement:

$$\left(\frac{\gamma}{\alpha}\right)_{\nu q} = \left(\frac{\gamma}{\alpha}\right)_{eD+\text{APV}} . \tag{7.2}$$

$$-0.450 \pm 0.206 \qquad -0.213 \pm 0.060$$

Here, the neutrino nucleon data lack precision which is due to the cancellation of the vector couplings for the u and d quark in the constant γ. The third model-independent test relates the v/a ratio from νe scattering to v/a from eD (see (5.4, 7)) with the quark couplings from νq scattering, thus factorising out the quark couplings:

$$\left(\frac{v}{a}\right)_{\nu e} = \left(\frac{a_2}{a_1}\right)_{eD} \left(\frac{2v_u - v_d}{2a_u - a_d}\right)_{\nu q} . \tag{7.3}$$

$$-0.077 \pm 0.095 \qquad 0.272 \pm 0.457$$

In this case the test is insignificant due to the large error on a_2 from polarised eD scattering, and again, substantial violations of factorisation are possible.

The conclusion to be drawn is that factorisation is a property of the standard theory which is not yet tested well. Note, however, that even in the standard theory small violations of factorisation are expected due to higher order weak corrections. At this stage, similar to the universality tests, no clear violations of the factorisation relations exist. This, again, may be evaluated as circumstantial evidence for the single Z^0 hypothesis as realized in the $SU(2) \times U(1)$ theory.

7.3 Weak Isospin Structure

According to the standard theory left-handed fermions f_L form weak isospin doublets while right-handed fermions f_R are singlets. There is strong evidence for the left-handed doublet structure, also for the third generation (see Chap. 3): Measurements on decays and lifetime of the τ lepton and the b quark clearly suggest partners not identical with any fermion from the first two generations. Both the τ neutrino and the t quark await positive discovery.

By analysing the axial-vector couplings one can test the singlet hypothesis for the righthanded fermions. As an example let us determine the value of the third component $t_3(l_R)$ of weak isospin for the right-handed leptons: One may use the values for a_l to calculate $t_3(l_R)$ with the help of (2.20) and (2.36):

$$a_l = 2 \left[t_3(l_L) - t_3(l_R) \right] . \tag{7.4}$$

With the experimentally verified weak isospin assignment $t_3(l_L) = -1/2$ for the left-handed leptons one obtains

$$t_3(e_R) = 0.005 \pm 0.026$$

$$t_3(\mu_R) = 0.022 \pm 0.028 \tag{7.5}$$

$$t_3(\tau_R) = -0.028 \pm 0.047 .$$

which clearly indicates that both right-handed muons and taus, like the electron, are singlets of weak isospin.

In the same way the third component of isospin for the right-handed quarks can be determined. With the usual isospin assignment for the left-handed fermions, $t_3(f_R) \geq 1/2$ is excluded to better than the 95 % C.L. for all fermions (see Table 7.1).

7.4 The Values of ϱ and $\sin^2 \theta_W$

The relative strength of the weak neutral current to the charged current is characterised by the parameter ϱ. In any $SU(2) \times U(1)$ model with spontaneous symmetry breaking ϱ is a function of the weak isospin of the Higgs fields (see (2.38)). In the special case of the standard theory with Higgs doublets, ϱ is expected to be equal to 1. The experimental value of ϱ is most precisely determined by deep inelastic ν scattering. In a global GSW fit to these data, including also results from the other available data on neutral current reactions, Amaldi et al. (1987) have obtained

$$\varrho = 0.998 \pm 0.009$$

$$\sin^2 \theta_W = 0.229 \pm 0.007 \tag{7.6}$$

in excellent agreement with the Higgs doublet assignment. $\varrho = 1$ is also well supported through the direct measurement of the W and Z masses, where the average over over the UA experiments yields

$$\varrho = 1.004 \pm 0.022 \qquad \bar{p}p \text{ only} .$$

When the value of ϱ is extracted from observables other than the weak boson masses, radiative corrections have to be included in principle, since the expectation $\varrho = 1$ is only strictly true in the lowest order of electroweak interactions. Loop level diagrams have to be taken into account which are calculable due to the renormalisability of the standard theory. The size of the higher order corrections depends on the momentum transfer Q^2 involved and on the particular physical process considered. Radiative corrections to $O(G\alpha)$ have been calculated for deep inelastic ν scattering by Marciano and Sirlin, (1981), which changes the observable ϱ to

$$\varrho(\nu q, \text{rad. corr.}) = 0.991 .$$

Such a small correction to ϱ is clearly unmeasureable at present. However, ϱ and $\sin^2 \theta_W$ are strongly correlated, as an elementary error analysis (Beg and Sirlin, 1982) shows. The deviation of ϱ from unity, due to the neglect of radiative corrections, induces a shift in the experimentally extracted value of $\sin^2 \theta_W$ of

$$\Delta(\sin^2 \theta_W) = 0.49 \, (\varrho^2 - 1) .$$

Thus for $\sin^2 \theta_W \sim 0.25$ a change of ϱ by -1% means a -4% change in $\sin^2 \theta_W$. Aiming at high precision measurements of $\sin^2 \theta_W$ thus strongly suggests the inclusion of radiative corrections.

One of the promising areas to test the standard theory at the one-loop level are the masses of the weak bosons. Recalling the discussion in Sect. 6.3, the radiative correction $\Delta \tilde{r}$ to the mass formulae (6.13), calculated from the boson masses but using the result on $\sin^2 \theta_W$ from neutrino nucleon scattering, was at the borderline of becoming significant. The significance can somewhat be improved by observing that the value for $\sin^2 \theta_W$ used in the calculation was already corrected for radiative effects. Following Amaldi et al. (1987) one should, when testing for radiative effects, also take out that part of the radiative correction and define the quantity δ_W:

$$\delta_W = \Delta \tilde{r} - \frac{\Delta s^2 (1 - \Delta \tilde{r})}{\sin^2 \theta_0} , \tag{7.7}$$

where $\sin^2 \theta_0$ is bare value of $\sin^2 \theta_W$ as determined from neutrino nucleon scattering (no radiative correction applied), and Δs^2 is the radiative correction. Using (7.7) one obtains

$$\delta_W = 0.122 \pm 0.33 , \tag{7.8}$$

which now exhibits a significance of the radiative correction at the almost 4σ level. When the statistical and systematic errors in deriving $\Delta \tilde{r}$ are added linearly, the error on δ_W is 0.043, still an almost 3σ effect. Fig. 7.1 shows the 90 % confidence level contours from the collider data in the $\Delta \tilde{r} - \sin^2 \theta_W$ plane (Amaldi et al., 1987). Also shown are the ordinates δ_W and $\sin^2 \theta_0$ used in the above analysis of the radiative corrections.

Although experimental precision is still not sufficient at present for most experiments to test the standard theory beyond the lowest order, it is still of

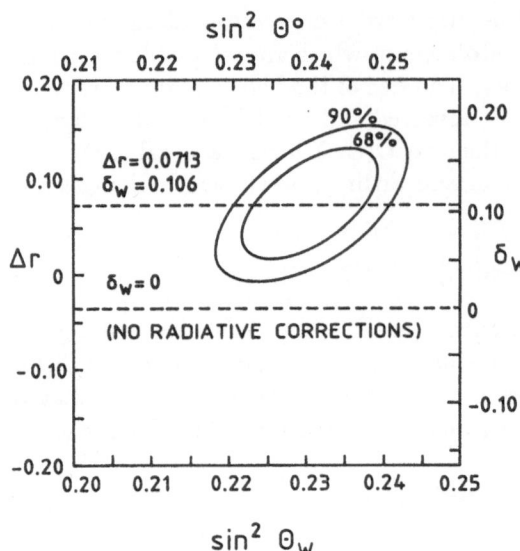

Fig. 7.1. Confidence level contours for $\sin^2 \theta_W$ vs. Δr from the $\bar{p}p$ experiments

$\sin^2 \Theta_W$

considerable interest to investigate the influence of higher order corrections on the determination of $\sin^2 \theta_W$ from the different measured processes. The corrections themselves, as stated above, depend on the process considered and on the kinematical region probed. Consistency among the renormalised values for $\sin^2 \theta_W$ from the different processes is highly non-trivial. The definition of the Weinberg angle depends on the renormalisation scheme used. In the on-mass-shell renormalisation scheme $\sin^2 \theta_W$ is defined by

$$\sin^2 \theta_W = 1 - M_W^2/M_Z^2 .$$

Other schemes, such as the modified minimal subtraction ($\overline{MS}$) scheme, lead to numerically very similar results for $\sin^2 \theta_W$ (Marciano, 1983), differing by less than a percent from the on-mass-shell definition. For a comprehensive review of the status of radiative corrections to weak processes and a detailed discussion of the various renormalisation procedures employed for calculating the Weinberg angle see, e.g., Beg and Sirlin (1982).

Radiative corrections have been calculated for all neutral current processes invested. The relevant calculations and references have been given in the preceeding chapters. As examples for the effect of the radiative corrections let us consider the two determinations from deep-inelastic neutrino nucleon scattering and polarised $e\,D$ scattering. The following shifts due to the radiative correction are observed for the values of $\sin^2 \theta_W$:

$$\nu q : \sin^2 \theta_W = 0.233 \pm 0.006 \quad (\text{ from } 0.243 \pm 0.006)$$

$$(7.9)$$

$$e D : \sin^2 \theta_W = 0.218 \pm 0.014 \quad (\text{ from } 0.224 \pm 0.014) .$$

Table 7.2. Determinations of the weak mixing angle

Exp.	$\sin^2 \theta_W$	Representative $Q^2[(\text{GeV}/c^2)^2]$
APV	0.217 ± 0.030	10^{-6}
$\nu\,e$	0.211 ± 0.021	10^{-2}
$e\,d$	0.218 ± 0.014	1
$\nu\,q$	0.233 ± 0.006	10
$\mu\,C$	0.230 ± 0.081	10^2
e^+e^-	0.217 ± 0.014	10^3
$\bar{p}\,p$	0.228 ± 0.008	10^4

Both corrections are negative and of about equal size thereby not destroying the agreement in the two experimental determinations. In fact, the radiative correction slightly improves the agreement between the two experiments.

In Table 7.2 the measurements of the Weinberg angle from the various neutral current reactions discussed in this report are compiled. Radiative corrections are applied in all cases. All available data on weak neutral interactions, spanning 10 orders of magnitude in Q^2, yield values for $\sin^2 \theta_W$ which agree with each other within about one standard deviation. The values for $\sin^2 \theta_W$, given in Table 7.2 are also shown in Fig. 7.2. All data are described accurately by just one parameter, the weak mixing angle $\sin^2 \theta_W$, a truely remarkable success of the theory. Taking the weighted average over all measurements in Table 7.2 one obtains

$$\sin^2 \theta_W = 0.228 \pm 0.004 \quad \text{(all data)}. \tag{7.10}$$

7.5 The Higgs Boson

In the standard theory neutral scalar (Higgs) bosons are inevitable for two reasons: they are responsible for the mass generation of the weak bosons by means of the symmetry breaking mechanism and they ensure the renormalisability of the theory by suitably choosing the couplings of the Higgs bosons to the fermions. This latter adjustment (Yukawa-type couplings) also gives rise to massive fermions. The masses of the fermions themselves, however, have to be introduced ad hoc, they are not calculable in the theory. The Higgs bosons have to be massive as well: Their fields assume non-zero vacuum expectation val-

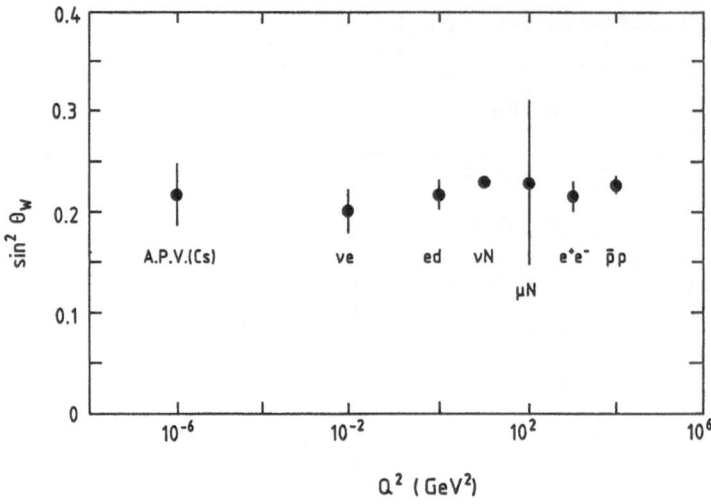

Fig. 7.2. Measurements of $\sin^2 \theta_W$ from the various experiments, characterized by their average momentum transfer Q^2. The values are corrected for radiative effects

ues, which constitutes the symmetry breaking mechanism. The limits on their masses, however, are not very stringent.

In lowest order, Higgs bosons do not contribute to the processes discussed above due to either vanishing (ν) or negigible (light fermion) couplings. Furthermore the quantum loop corrections mentioned in the previous paragraph show a very weak dependence on the Higgs mass (Veltman, 1977). Still, on rather general grounds, one can limit the range of allowed masses for the Higgs boson: A lower boundary for the Higgs mass m_H has been derived from the requirement of stability of the Higgs vacuum in presence of radiative corrections (Weinberg, 1976) yielding

$$ m_H > \frac{\alpha}{\sin^2 \theta_W} \frac{\sqrt{3}}{4} \left[\frac{2 + \cos^{-4} \theta_W}{\sqrt{2} G} \right]^{1/2} . $$

For $\sin^2 \theta_W \sim 0.23$ one obtains

$$ m_H > 6.8 \quad [\text{GeV}/c^2] . $$

It should be kept in mind, however, that in certain non-standard Higgs scenarios this lower limit is not a very solid one (see, e.g., Gunion, 1988).

In contrast, a more reliable upper limit can be found (Veltman, 1977a) by considering the hypothetical process $W^+W^- \to W^+W^-$ which can be mediated by γ, Z^0 and H exchange. The energy dependence of the cross section due to γ or Z exchange is proportional to s, which can be largely cancelled by the Higgs exchange diagram with an appropriate choice of m_H. In order to conserve unitarity in the $J = 0$ partial wave, one can deduce

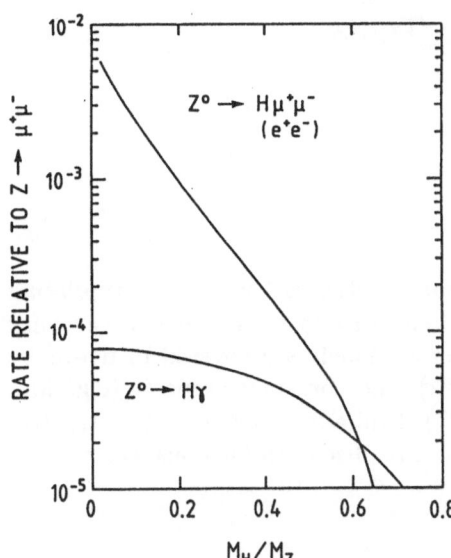

Fig. 7.3. Production rates of the Higgs particle at the Z^0 resonance, relative to $Z^0 \to \mu^+\mu^-$, for different final states as functions of the Higgs mass

$$m_{\mathrm{H}}^2 < \frac{8\pi\sqrt{2}}{G}$$

which means an upper limit around $1 - 2$ TeV$/c^2$.

Since the Higgs coupling to fermions is proportional to the fermion mass heavy quarkonium states may be searched for the decay

$$(Q\bar{Q}) \to \gamma \; H$$
$$\hookrightarrow f\bar{f}$$

The experimental upper limits for $(Q\bar{Q}) = J/\psi$ and Υ, however, are still much larger than the theoretical predictions (see Yamada (1983) for a review of Higgs searches). The most promising decays to discover the Higgs are $t\bar{t} \to \gamma H$ and $Z^0 \to \gamma H$, e^+e^-H, $\mu^+\mu^-H$. Figure 7.3 shows the expected production rate of Higgs particles at the Z^0 pole. Since the yield is steeply falling with the Higgs mass, even at the high luminosities expected at LEP the detection of heavy Higgses will not be easy. Bearing in mind that m_{H} can be as large as 1 TeV$/c^2$ it is conceivable that the Higgs particle may escape detection for quite some time.

8. Extensions and Further Tests of the Standard Theory

Literature abounds with models going beyond the standard theory or formulating alternatives, and it is certainly inappropriate here to try for an adequate account of the ingenuity put into the subject. These models are covered in depth by review articles such as Beg and Sirlin (1982), and conference reports (e.g. Ma, 1983; Barbieri, 1983; Ellis, 1983; Ross, 1987). Still a few theoretically attractive examples and their experimental tests should be discussed for illustration.

8.1 Beyond the Standard Model

After the discovery of the heavy bosons with precisely the masses predicted by the standard theory many of the non-gauge, non-unification models have lost credibility. Both electroweak and strong interactions (QCD) seem to be well described by gauge theories (see, e.g., Politzer, 1974; Marciano and Pagels, 1978; Mueller, 1981; Lepage, 1983; Duke and Roberts, 1987) with the presently favoured gauge group $SU(3)_C \times SU(2)_L \times U(1)$. An intriguing idea is the assumption that this group is embedded in a larger, simple group G with a single gauge coupling constant, thus achieving a fundamental unification of strong and electroweak forces at some (large) mass scale M_X. Below that mass scale, G is spontaneously broken into the above factor groups, leading to the evolution of very different coupling constants for the strong and electroweak interaction observed at energies small compared to M_X. The simplest possibility for a Grand Unified Theory (GUT) is to choose $SU(5)$ for the group G (Georgi and Glashow, 1974). In minimal (only one Higgs scalar) $SU(5)$ GUT no new physics is expected in the energy range between the mass of the weak bosons and the unification scale of order 10^{15} GeV, often called the "desert". Within $SU(5)$ the value of the unrenormalised Weinberg angle as defined in (2.15) is given by

$$\sin^2 \theta_W = 3/8 .$$

The result is independent of the number of quark/lepton generations. This value of $\sin^2 \theta_W$ would be measured if $SU(5)$ were an unbroken gauge symmetry. Detailed calculations of the evolution of coupling constants as one moves down from the unification point (Marciano and Sirlin, 1981a; Llewellyn Smith et al., 1981; Hall, 1981) yield the renormalised value of $\sin^2 \theta_W$ at present laboratory energies:

$$\sin^2 \theta_W = 0.216 + 0.006 \ln(0.1 \, \text{GeV}/\Lambda_{\overline{\text{MS}}}) \,,$$

where $\Lambda_{\overline{\text{MS}}}$ is the QCD scale parameter. For $\Lambda_{\overline{\text{MS}}}$ around 200 MeV, the preferred value from Υ decay where the most precise measurement has been made, and most other data (Stirling, 1987), this value is in tantalising vicinity to the radiatively corrected experimental results in (7.10). However, minimal $SU(5)$ GUT where quarks and leptons are found in the same representation of the group predicts a finite lifetime τ_p for the proton:

$$\tau_p \sim 10^{29 \pm 2} \quad \text{years}$$

in conflict with recent experimental lower limits (Bionta et al., 1983; Meyer, 1986)

$$\tau(p \to e^+ \pi^0) > 4 \times 10^{32} \quad \text{years} \,. \qquad (90\% \,\text{C.L.})$$

No single event has been found in the various nucleon decay experiments over the past years which is undoubtedly stemming from proton decay. Since the minimal $SU(5)$ model seems to be in disagreement with the data the significance of the agreement between the renormalised theoretical and experimental values of $\sin^2 \theta_W$ remains unclear.

Why would one worry and think about possible extensions beyond the standard theory at all? The answer lies in the fact that the standard theory, despite all its sucesses, cannot be the ultimate description of Nature. Too many fundamental questions remain unanswered and too many arbitrary parameters are involved in the theory. Let us mention just a few of the problems:

There is no explanation concerning the number of families. Why are there three generations, how many are there?

The fermion masses enter as free parameters into the theory. No explanation of the mass spectrum is given, neither is there any clue as to why the quarks are mixed in such complicated multiplets with a large number of unknown mixing angles.

There is no recipe at present to include gravitation into a unified interaction. Gravity can be put in only "by hand".

This latter aspect of incoporating gravity has led to the concept of supersymmetry, where, from a purely phenomenological view point, a new class of scalar particles must exist, each associated with its fermionic partner, the masses of which are not known a priori.

Another approach is the compositeness model, where the quarks and leptons are made of even more elementary building blocks. Questions such as the family problem can be addressed, at least in principle.

In most cases, the new theoretical ideas lead to experimental consequences. In all cases more detailed and precise experimental tests of the standard theory and possible extensions are needed.

All extensions or alternatives to the standard theory are severely constrained in their low energy predictions: They have to reproduce the standard theory for

$\sqrt{s} \leq M_Z$. For energies at or well above the weak boson masses experimental data are sparse giving still a lot of freedom to model builders. Many models as the ones discussed above with gauge groups of the type $G \times SU(2) \times U(1)$ or more general mixing schemes, lead to the following effective neutral current Lagrangian in the low energy limit (Kuroda and Schildknecht, 1982):

$$\mathcal{L}_{\text{eff}}^{\text{NC}} = -\frac{4G}{\sqrt{2}} \left[\left(J_\lambda^3 - \sin^2 \theta_W J_\lambda^{\text{em}} \right)^2 + C \left(J_\lambda^{\text{em}} \right)^2 \right] .$$

This Lagrangian differs from (2.30) by a term proportional to the square of the electromagnetic current. Since J_λ^{em} conserves parity, such a contribution cannot be found in ν scattering or other parity-violating processes like polarised $e\,d$ scattering. It modifies, however, the square of the vector coupling constant in e^+e^- experiments in the following way:

$$v^2 = (-1 + 4 \sin^2 \theta_W)^2 + 16\,C .$$

Taking the value of $\sin^2 \theta_W$ from the parity violating experiments ($\sin^2 \theta_W = 0.229 \pm 0.005$), one can derive limits for C using the value for v^2 from all e^+e^- experiments, given in (3.57). One obtains

$$C < 0.028 \quad \text{at the 95 \% C.L.} .$$

Within specific models, C constrains the masses and couplings of additional weak neutral bosons heavier than the Z^0.

8.2 Precise Tests

Further tests of the standard theory are certainly needed, both on the qualitative and quantitative level.

Positive evidence is needed for the τ neutrino, the top quark and the Higgs boson. The τ neutrino may ultimately be found in beam dump experiments while the top quark and possibly the Higgs particle may need the next generation of accelerators to be identified (the Tevatron Collider, SLC, LEP, TRISTAN, HERA).

The spectrum of particles needs further clarification, in particular the number of lepton/quark families. Searches for new sequential fermions have been unsuccessful so far. As an example, the mass limits obtained for new heavy neutral leptons are shown in Fig. 8.1, which could exhibit mixing similar to the quarks. For a recent review of new particle searches see Kamae (1988). Furthermore the width of the Z^0 provides information about the number of generations of quarks and leptons realised in nature: In principle, one can use a direct measurement of Γ_Z as a bound for the maximum number of light neutrinos, assuming that all other decay modes are known. In practice one tries to estimate Γ_Z using the

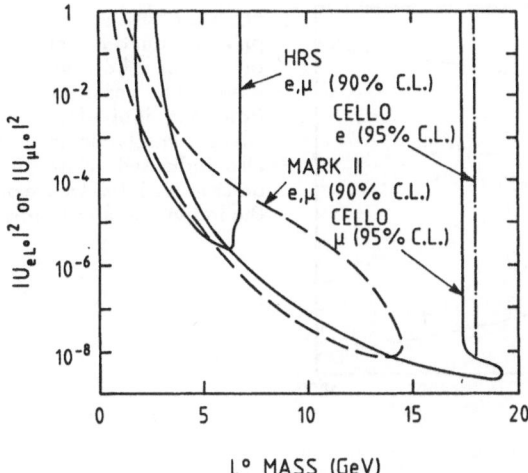

Fig. 8.1. Limits for the production of new neutral leptons from e^+e^- experiments, as a function of their couplings to electrons or muons, compiled by Kamae (1988)

following expression:

$$\cdot \Gamma_Z{}' = \Gamma_W \left(\frac{\sigma_Z}{\sigma_W} \right) \cdot \left(\frac{\Gamma_e^+ e^-}{\Gamma_{e\nu}} \right) \cdot R \,,$$

where the quantity R, defined as

$$R = \frac{\sigma_W \cdot \mathrm{BR}(W \to e\nu)}{\sigma_Z \cdot \mathrm{BR}(Z^0 \to e^+e^-)}$$

is measured in the experiment and the values for $\Gamma_W, \Gamma_e^+ e^-$ and $\Gamma_{e\nu}$ are assumed according to the standard theory. The cross section ratio σ_W/σ_Z is calculated using a QCD model. With this analysis the combined UA value for the number of light neutrinos is (Colas et al., 1988)

$$N_\nu < 5.7 \quad \text{at 90\% C.L.} \quad (\bar{p}p) \,.$$

In a recent analysis CELLO (Behrend et al., 1988) have searched for single photon events resulting from the reaction

$$e^+e^- \to \gamma\nu\bar{\nu} \,,$$

where the production of a neutrino pair is tagged by a photon radiated in the initial state. CELLO have combined their results with other e^+e^- neutrino counting experiments and obtain

$$N_\nu < 4.6 \quad \text{at 90\% C.L.} \quad (e^+e^-) \,.$$

It is interesting to note that astrophysical limits on light neutrinos, based on nucleosynthesis and supernova observations (Steigman et al., 1986; Schaeffer et

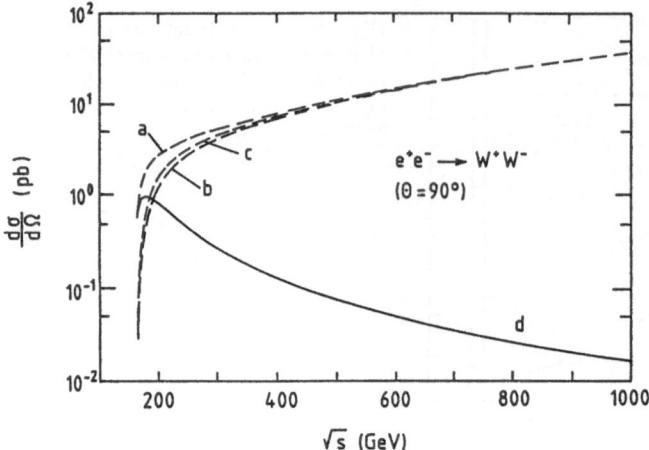

Fig. 8.2. Cross section for pair production of W bosons in e^+e^- annihilation. The contributions from the individual exchange mechanisms are also indicated. Note the destructive interference of the individual amplitudes

al., 1987) lead to quite similar results. There seems agreement among these searches that the number of generations does not exceed four or, at most, five.

High precision measurements of the masses of the weak bosons W and Z are needed to test the quantum loop corrections to the tree level calculations. Considering the mass difference $\Delta = M_Z - M_W$, the loop corrections may even give a hint to the mass m_t of the top quark. For a t quark mass comparable to M_W the quantity Δ is given by (Llewellyn Smith, 1983)

$$\Delta \simeq \left(10.6 - 0.14\,\frac{m_t^2}{M_W^2}\right)[\mathrm{GeV}]\,.$$

Thus loop corrections tend to diminish the mass difference while $\Delta > 10.6$ GeV may indicate new physics. Composite models, e.g., might produce a mass difference in excess of 11 GeV (Schildknecht, 1983; Fritzsch, 1983). Precise measurements ($O(1\%)$) of the mass difference may be obtained within the next few years at the e^+e^- and $\bar{p}p$ colliders.

A very interesting and unique prediction of any non-abelian gauge theory is the self-coupling of the gauge bosons. The cleanest way to test the $Z^0 - W^\pm$ coupling is certainly the reaction $e^+e^- \to W^+W^-$ which will be accessible at LEP II. Figure 8.2 shows the cross section for W pair production. The individual amplitudes contributing are of the same order of magnitude so that the negative interference between the Z^0 and ν exchange amplitudes produces a very clear effect.

Indications that the "great desert" might be populated are provided by the apparent failure of minimal $SU(5)$ GUT. A richer particle spectrum beyond the weak bosons may also help to dynamically explain a number of ad hoc assumptions of the standard theory, such as the weak multiplet structure, i.e. parity violation, the number of fermion generations, the fermion-Higgs couplings, fermion masses etc.. Straightforward extensions of the standard theory to restore parity invariance are right-left symmetric models (Pati and Salam, 1974; Mohapatra

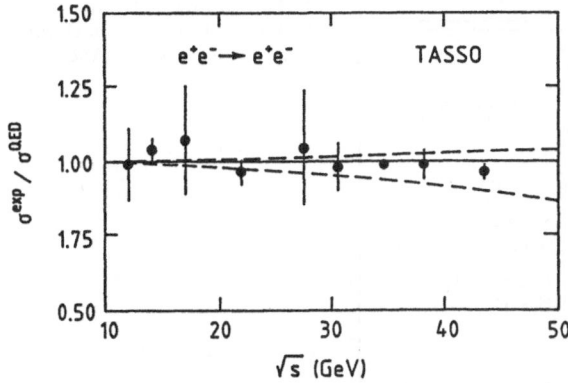

Fig. 8.3. Limits of compositeness from the differential cross section for Bhabha scattering integrated over $|\cos\theta| < 0.8$ (TASSO). The dashed curves show the deviations from QED for cut-off parameters $\Lambda_+ = 370\,\mathrm{GeV}$ and $\Lambda_- = 190\,\mathrm{GeV}$

and Senjanovich, 1980, 1981). The observed parity violation at low energies is explained by allowing for a sufficiently heavy boson W_R coupling to right-handed $(V + A)$ currents. Precise measurements of the endpoint of the β spectrum from polarised muon decay (Carr et al., 1983) have set the limit $M(W_\mathrm{R}) > 380$ GeV, assuming light $(m(\nu_\mathrm{R}) < 10\,\mathrm{MeV})$ right-handed neutrinos. Relaxing the ν mass condition lowers the limit considerably $(M(W_\mathrm{R}) > 200\,\mathrm{GeV})$. Measurements at the future $e\,p$ machine HERA providing longitudinal polarisation of the electron will improve this limit somewhat. In more restrictive left-right symmetric models (Beall et al., 1982; Branco et al., 1983), analysing the $K_L^0 - K_S^0$ mass difference in the framework of a four-quark model, rather strong bounds, $M(W_\mathrm{R}) \geq 1.5$ TeV, are obtained.

Another way to populate the desert, and at the same time address the question of proliferation of "elementary" particles, is found in composite models (for a review see, e.g., Peskin, 1980; Ross, 1987) where compositeness may refer to Higgs particles, leptons and quarks or even the gauge bosons. A lower limit for the compositeness scale Λ in the case of leptons may be obtained from their anomalous magnetic moment. Using Λ as a cut-off in the muon form factor, a bound of $\Lambda > 800$ GeV results. Somewhat lower limits are obtained for electrons, where Bhabha scattering can be used to search for contributions from additional contact terms to the standard electroweak cross section, characterised by the compositeness scales $\Lambda_\pm$. Figure 8.3 shows the Bhabha scattering cross section from TASSO (Braunschweig et al., 1988), normalised to the lowest order QED expectation, as a function of the centre of mass energy. From these data the 90 % C.L. limits $\Lambda_+ > 435, \Lambda_- > 590$ GeV are derived.

Recently the potentially composite nature of the weak bosons has been investigated in the literature (Fritzsch, 1983; Fritzsch, Schildknecht and Kögerler, 1982; Peccei, 1984;), triggered by the possible observation of Z^0 decays into $e^+e^-\gamma$ at the $\bar{p}p$ collider. A particular model (Peccei, 1984) interpreting $Z \to e^+e^-\gamma$ as a dipole transition to a scalar / pseudoscalar particle X, subsequently decaying to e^+e^- has been excluded by the CELLO collaboration at PETRA (Behrend et al., 1984a) searching for X exchange contributions in $e^+e^- \to e^+e^-, \mu^+\mu^-$ and $\tau^+\tau^-$. With the full statistics from the UA experiments anal-

191

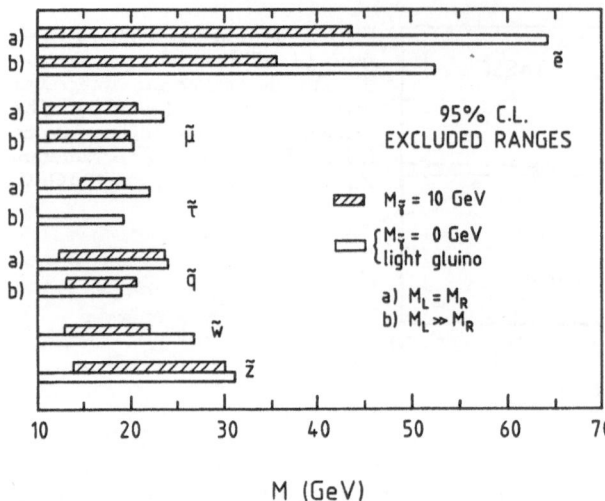

Fig. 8.4. Limits for the mass of supersymmetric particles, compiled by Kamae (1988)

ysed the two events found are consistent with the expectation from internal Bremsstrahlung (Ansari et al., 1987).

A most attractive scheme from a theoretical point of view, with the potential of unifying strong, electroweak and even gravitational forces is provided by supersymmetric (SUSY) theories (Gelfand and Likhtman, 1971; Volkov and Akulov, 1973; Wess and Zumino, 1974, and Fayet and Ferrara, 1977). New particles are predicted which are the scalar partners of the known fermions. Their mass scale is unknown, but could be well below 1 TeV. SUSY theories have the general property to raise the unification scale thereby reconciling with the problem of the nucleon lifetime, while retaining the predictive-power for the renormalised Weinberg angle (Ellis et al., 1982). Most collider (e^+e^- and $\bar{p}p$) experiments have searched for such particles, which have specific and clearly discriminating signatures, none with success. The present status of SUSY searches is shown in Fig. 8.4, including the new results from TRISTAN (Kamae, 1988). Again, the next generation of accelerators or more careful and precise experiments may bring new insight.

No experimental data exist so far which are in conflict with the standard theory. Nothing hints towards a new class of particles which necessarily calls for an extension of the theory. But too many questions await an answer which are not addressed by the standard theory. It thus seems very unlikely that the correct theory has been found. Probing the > 0.1 TeV region with the $\bar{p}p$ colliders or with the next generation of e^+e^- and ep machines will certainly be very exciting.

9. Conclusion

The standard theory of electroweak interactions, as formulated by Glashow, Salam and Weinberg, based on the gauge group $SU(2) \times U(1)$, seems to be firmly established. All experiments carried out so far are quantitatively described by this simplest possible unification scheme of the electromagnetic and the weak interaction, introducing one additional parameter, the Weinberg angle, usually given as $\sin^2 \theta_W$. The most dramatic prediction, the existence of heavy gauge bosons $W^\pm$ and Z^0, has been verified by experiment.

Measurements of the Weinberg angle from the various experiments, spanning the range from atomic parity violation with momentum transfers Q^2 of order 10^{-6} GeV^2 up to the production threshold for the weak bosons, with Q^2 of order 10^4 GeV^2, are in excellent agreement. Moreover, the radiative corrections which are applied to the different experiments bring the $\sin^2 \theta_W$ determinations in even better accord. Combining the measurements of the boson masses with results on $\sin^2 \theta_W$ from neutrino nucleon scattering lead to a significant test of the theory at the one-loop level. Also here the predictions of the theory are in excellent agreement with the measurements.

Fermion pair production from e^+e^- has contributed substantially to test various predictions and assumptions of the standard theory: The wealth of data from PEP, PETRA and TRISTAN on leptons and quarks of the second and third generation have convincingly demonstrated the universal character of the electroweak interaction. The precision of the $\sin^2 \theta_W$ determinations from e^+e^- experiments is comparable now to the classical experiments in deep-inelastic scattering.

Measurements of the relative strength of neutral to charged current in neutrino nucleon reactions and the mass ratio of the weak bosons have clearly shown that the symmetry-breaking Higgs mechanism, one of the corner stones of the standard theory, seems to be realised, again in its simplest form as suggested by GSW. The lesson to be learned from these findings probably is that local gauge theories with spontaneous symmetry breaking are Nature's recipe to generate forces amongst elementary particles.

Despite all its successes the standard theory is certainly not complete, and therefore not the ultimate theory of particle interactions. Many free parameters are needed to make any prediction in the standard theory, most importantly the masses of the fermions. There is no clue in the standard theory as to why the particle spectrum is as it is. In the charged current sector, not discussed in this review, there is no explanation why the quarks mix and why the leptons do not. Again a number of free parameters, the elements of the CKM matrix, have to be introduced. Finally there is no answer to the family question. Why do we have more than one familiy, how many families are there anyway?

One of the sad predictions, in the most naive scenario of the standard theory, is the "great desert" beyond the Z^0 pole. In the quest to solve or at least address some of the open questions theorists have populated the desert, with lots of species waiting to be discovered by the keen experimenter. The physical origin of these new particles is completely open and left to anybody's imagination. Promising speculations are put forward in composite models where the Higgs scalars, fermions, and even gauge bosons may be built from more elementary constituents. Supersymmetric models bring us even more new particles, advocating a rich spectrum of scalar partners for the known fermions at some large mass scale.

After two decades of spectacular success theory is now somewhat in disarray. Experimental hints favouring a specific direction of thought are missing. Agreement, however, seems to exist amongst most theoretical ideas proposed that the 1 TeV scale will reveal the underlying scheme leading to present-day phenomenology. Maybe, some of the known elementary particles are really composite and the weak force, at the moment the last fundamental interaction with finite range, is only a van der Waals remnant of a new kind of hypercolour force, binding fermionic and/or bosonic constituents together. Speculation is abundant − experiments probing the deeper structure of matter at a new energy scale are called for.

References

Abachi, S. et al. (1985): Preprint ANL-HEP-PR-85-113

Abbott, L.F., R.M. Barnett (1978): Phys. Rev. **D18** 3214

Abe, K. et al. (1986): Phys. Rev. Lett. **56** 1107

Abrams, G.S. et al. (1982): Phys. Rev. Lett. **48** 1586

Abramovicz, H. et al. (1981): Phys. Lett. **109B** 115

Abramovicz, H. et al. (1985): Z. Phys. **C28** 51

Abramovicz, H. et al. (1986): Phys. Rev. Lett. **57** 298

Achenbach, W. et al. (1986): Proceedings of the International Symposium on Weak and Electromagnetic Interactions in Nuclei, Heidelberg, 1986, p. 642

Adeva, B. et al. (1982): Phys. Rev. Lett. **48** 1701

Adeva, B. et al. (1983): Phys. Rev. Lett. **50** 799

Adeva, B. et al. (1983a): Phys. Rev. Lett. **51** 443

Adeva, B. et al. (1984): Phys. Rep. **109** 131

Adeva, B. et al. (1984a): Phys. Lett. **152B** 297

Adeva, B. et al. (1985): Phys. Rev. Lett. **55** 665

Adeva, B. et al. (1986): Phys. Lett. **179 B** 177

Adeva, B. et al. (1986a): Phys. Rev. **D34** 681

Ahlen, S. et al. (1983): ANL Preprint ANL-HEP-PR-83-39

Ahlen, S. et al. (1983a): Contribution no. 157 to the Lepton Photon Conference Cornell (1983)

Ahlen, S. et al. (1983b): Phys. Rev. Lett. **51** 1147

Ahrens, L.A. et al. (1983): Phys. Rev. Lett. **51** 1514

Ahrens, L.A. et al. (1985): Phys. Rev. Lett. **54** 18

Ahrens, L.A. et al. (1987): Phys. Rev. **D35** 785

Aihara, H. et al. (1983): IEEE Transactions on Nuclear Science, Vol NS-30, No 1, p. 63, 67, 76, 117, 153

Aihara, H. et al. (1984): Phys. Rev. **D30** 2436

Aihara, H. et al. (1985): Phys. Rev. **D31** 2719; Z. Phys. **C27** 39

Aihara, H. et al. (1986): Phys. Rev. **D34** 1945

Aitchison, I.J.R., A.J.G. Hey (1982): Gauge Theories in Particle Physics, Adam Hilger Ltd, Bristol (1982):

Akerlof, C. et al. (1985): Phys. Rev. Lett. **55** 570

Albajar, C. et al. (1987): CERN -EP/87-149, submitted to Phys. Lett. B

Albrecht, H. et al. (1985): Phys. Lett. **163B** 404

Albrecht, H. et al. (1985a): Phys. Lett. **150B** 235

Albrecht, H. et al. (1987): DESY preprint 87-148

Albrecht, H. et al. (1987a): Phys. Lett. **192B** 245

Alexander, G. et al. (1978): Phys. Lett. **78B** 162

Allaby, J. V. et al. (1986): Phys. Lett. **177B** 446

Allasia, D. et al. (1983): Phys. Lett. **133B** 129

Allen, R.C. et al. (1985): Phys. Rev. Lett. **55** 2401

Altarelli, G., R.K. Ellis, M. Greco and G. Martinelli (1984): Nucl. Phys. **B246** 12

Altarelli, G., R.K. Ellis, M. Greco and G. Martinelli (1985): Z. Phys. **C27** 617

Althoff, M. et al. (1983): Phys. Lett. **126B** 493

Althoff, M. et al. (1984): Z. Phys. **C22** 13

Althoff, M. et al. (1984a): Phys. Lett. **138B** 441

Althoff, M. et al. (1984b): Phys. Lett. **141B** 264

Althoff, M. et al. (1985): DESY - Preprint 85-001, Jan. 1985, and TASSO Note 339 (unpublished)

Althoff, M. et al. (1985a): Z. Phys. **C26** 521

Amaldi, U. et al. (1987): Phys. Rev. **D36** 1385

Ammosov, V. V. et al. (1983): Proceedings of the 21st International Conference on High Energy Physics, Paris 1982

Ammosov, V. V. et al. (1986): Z. Phys. **C30** 569

Ammosov, V. V. et al. (1986a): preprint PHE 86-10

Anderson, B., G. Gustavson, T. Sjöstrand (1980): Phys. Lett. **94B** 211

Ansari, R. et al. (1987): Phys. Lett. **186B** 440; **190B** 238E

Appel, J.A. et al. (1985): CERN-EP / 85-166, submitted to Z. Phys. C

Apperson, G.A. (1979): Ph.D. Thesis , University of Washington, Seattle (unpublished)

Argento, A. et al. (1983): Phys. Lett. **120B** 245

Argento, A. et al. (1984): Phys. Lett. **140B** 142

Armenise, N. et al. (1979): Phys. Lett. **81B** 385

Armenise, N. et al. (1979a): Phys. Lett. **86B** 255

Armenise, N. et al. (1983): Phys. Lett. **122B** 448

Arnison, G. et al. (1983a): Phys. Lett. **122B** 103

Arnison, G. et al. (1983b): Phys. Lett. **126B** 398

Arnison, G. et al. (1983c): Phys. Lett. **129B** 273

Arnison, G. et al. (1984d): Phys. Lett. **134B** 469

Arnison, G. et al. (1985): CERN-EP / 85-185, Europhys. Lett. **1** (1986) 327; Phys. Lett. **166B** (1986) 484

Ash, W.W. et al. (1985): SLAC-PUB-3741, July 1985

Astratyan, A.E. et al. (1979): Proceedings of the International Conference on Neutrino Physics '79, Bergen, p. 246

Atwood, W.B. et al. (1983): SLAC-PUB-3260

Aubert, J.J. et al. (1974): Phys. Rev. Lett. **33** 1404

Aubert, J.J. et al. (1983): Phys. Lett. **123B** 123

Augustin, J.E. et al. (1974): Phys. Rev. Lett. **33** 1406

Augustin, J.E. (1979): Proceeding of the LEP Summer Study, CERN 79-01, p. 499

Avery, P. et al. (1983): Phys. Rev. Lett. **51** 1139

Avignone, F. T., Z. D. Greenwood (1977): Phys. Rev. **D16** 2383

Bacino, W. et al. (1978): Phys. Rev. Lett. **40** 671

Bacino, W. et al. (1979): Phys. Rev. Lett. **42** 6

Bagnaia, P. et al. (1983): Phys. Lett. **129** 130

Bagnaia, P. et al. (1984): Z. Phys. **C24** 1

Baird, P.E.G. et al. (1977): Phys. Rev. Lett. **39** 798

Baird, P.E.G. et al. (1980): Proceedings of the International Workshop on Neutral Current Interactions in Atoms, Cargese , p. 77

Baker, N.J. et al. (1983): BNL preprint BNL-33572, Sept. 1983

Baltay, C. et al. (1986): Phys. Rev. Lett. **57** 2629

Banner, M. et al. (1983): Phys. Lett. **122B** 476

Baranko, G. et al. (1985): Proceedings of the International Europhysics Conference on High-Energy Physics, Bari, p. 226

Barber, D.P. et al. (1980): Physics Reports **63** 337

Barbiellini, G., C. Santoni (1986): Riv. Nuo. Cim. **9** 1

Barbieri, R. (1983): Proceedings of the International Lepton - Photon Symposium, Cornell University, Ithaca , p. 479

Bardeen, W. A. et al. (1978): Phys. Rev. **D18** 3998

Bardin, D.Yu. et al. (1980): Yad. Fiz. **32** 782

Bardin, D.Yu. et al. (1982): Nucl. Phys. **B197** 1

Bardin, et al. (1986): Preprint JINR-E2-86-260; Sov. Yad. Phys. **30** (1979) 811; Sov. Yad. Phys. **36** (1982) 482

Baringer, P. et al. (1988) Phys. Lett. **B206** 551

Barkov, L., M. Zolotorev (1978): JETP Lett. **27** 379

Barkov, L., M. Zolotorev (1979): Phys. Lett. **B48** 524

Barkov, L.M. (1982): Proc. Neutrino-82 Conference, Balatonfüred, Budapest 1982, Vol. II, p. 89

Barnett, R.M. (1976): Phys. Rev. **D14** 70

Bartel, W. et al. (1978): Phys. Lett. **77B** 331

Bartel, W. et al. (1979): Phys. Lett. **88B** 171

Bartel, W. et al. (1981): Phys. Lett. **100B** 364

Bartel, W. et al. (1982): Phys. Lett. **108B** 140

Bartel, W. et al. (1983): Z. Phys. **C19** 197

Bartel, W. et al. (1983a): DESY 83 - 049

Bartel, W. et al. (1983b): Phys. Lett. **129B** 145

Bartel, W. et al. (1984): Phys. Lett. **146B** 121

Bartel, W. et al. (1984a): Phys. Lett. **146B** 437

Bartel, W. et al. (1985): DESY preprint 85-131

Bartel, W. et al. (1985a): Phys. Lett. **160B** 337

Bartel, W. et al. (1985b): Phys. Lett. **161B** 188

Bartel, W. et al. (1987): Contribution to the 1987 International Symposium on Lepton and Photon Interactions at High Energies, Hamburg

Batley, R. (1985): Proceedings of the 6th Workshop on $\bar{p}p$ Collider Physics, Aachen, July 1985

Beall, G., M. Bander, A. Soni (1982): Phys. Rev. Lett. **48** 848

Bebek, C. et al. (1982): Phys. Rev. Lett. **49** 610

Beg, M.A.B., A. Sirlin (1982): Phys. Rep. **88** 1

Behrend, H.J. et al. (1981): Physica Scripta **23** 610
Behrend, H.J. et al. (1981a): Phys. Lett. **103B** 148
Behrend, H.J. et al. (1981b): DESY 81-029
Behrend, H.J. et al. (1982): Z. Phys. **C14** 283
Behrend, H.J. et al. (1982a): Z. Phys. **C14** 283
Behrend, H.J. et al. (1982b): Phys. Lett. **114B** 282
Behrend, H.J. et al. (1983): Z. Phys. **C16** 301
Behrend, H.J. et al. (1983a): Phys. Lett. **127B** 270
Behrend, H.J. et al. (1983b): Nucl. Phys. **B211** 369
Behrend, H.J. et al. (1983c): Z. Phys. **C 19** 291
Behrend, H.J. et al. (1983d): Nucl. Phys. **B218** 269
Behrend, H.J. et al. (1984): Z. Phys. **C23** 103
Behrend, H.J. et al. (1984a): Phys. Lett. **140B** 130
Behrend, H.J. et al. (1984b): DESY Preprint 84-020
Behrend, H.J. et al. (1984c): Phys. Lett. **138B** 311
Behrend, H.J. et al. (1984d): Phys. Lett. **144B** 297
Behrend, H.J. et al. (1985): Contribution to Bari, Kyoto
Behrend, H.J. et al. (1987): Phys. Lett. **B183** 400
Behrend, H.J. et al. (1988): DESY preprint 88-052, to be published in Phys. Lett.
Beltrami, I. et al. (1985): Phys. Rev. Lett. **54** 1775
Bender, D. et al. (1985): Phys. Rev. **D31** 1
Berends, F.A. et al. (1974): Nucl. Phys. **B68** 541
Berends, F.A., G.J. Komen (1976): Phys. Lett. **63B** 432
Berends, F.A., R. Kleiss (1981): Nucl. Phys. **B177** 237
Berends, F.A., R. Kleiss (1981a): Nucl. Phys. **B178** 141
Berends, F.A. et al. (1982): Nucl. Phys. **B202** 63
Berends, F.A., R. Kleiss (1983): Nucl. Phys. **B228** 537
Berge, J.P. et al. (1979): FNAL preprint 79/27
Berger, Ch. et al. (1980): Phys. Lett. **91B** 148
Berger, Ch. et al. (1981): Phys. Lett. **99B** 489
Berger, Ch. et al. (1983): Z. Phys. **C21** 53
Berger, Ch. et al. (1985): Z. Phys. **C27** 341
Berger, Ch. et al. (1985a): Z. Phys. **C28** 1
Bergsma, F. et al. (1982): Phys. Lett. **117B** 272
Bergsma, F. et al. (1984): Phys. Lett. **147B** 481
Bergsma, F. et al. (1985): Phys. Lett. **157B** 469
Berman, S.M. and J.R. Primack (1974): Phys. Rev. **D9** 2171; Phys. Rev. **D10** 3895 (E)
Bertrand-Coremans, G. et al. (1976): Phys. Lett. **61B** 207
Bethke, S. (1985): DESY Preprint 85-067, July 1985
Bilenky, S.M. et al. (1977): Phys. Lett. **67B** 309
Bionta, R.M. et al. (1983): Phys. Rev. Lett. **51** 27
Bjorken, J.D. (1976): Proceedings of the Summer Institute on Particle Physics, SLAC, Stanford 1976 p. 1
Bjorken, J.D. (1978): Phys. Rev. **D17** 171

Bjorken, J.D. (1978a): Phys. Rev. **D18** 3239

Blair, R.E. et al. (1983): Fermilab Preprint 83/76-Exp 1983

Blietschau, J. et al. (1978): Phys. Lett. **73B** 232

Blietschau, J. et al. (1979): Phys. Lett. **87B** 281

Blietschau, J. et al. (1979a): Phys. Lett. **88B** 381

Blocker, C.A. et al. (1982a): Phys. Lett. **109B** 119

Blocker, C.A. et al. (1982b): Phys. Rev. Lett. **49** 1369

Bludman, S.A. (1958): Nuov. Cim. **9** 443

Böhm, M., W. Hollik (1984): Phys. Lett. **139B** 213

Bodek, A. et al. (1979): Phys. Rev. **D20** 1471

Bogdanov, Yu. V. et al. (1980): JETP Lett. **31** 522

Bogert, D. et al. (1985): Phys. Rev. Lett. **55** 1969

Bonneau, G., A. Martin (1971): Nucl. Phys. **B27** 381

Booth, C. (1986): Proceedings of the 6th Topical Workshop on Proton-Antiproton Collider Physics, Aachen, 1986, p.381

Bosetti, P. et al. (1983): Nucl. Phys. **B217** 1

Bouchiat, M.A., C. Bouchiat (1974): Phys. Lett. **48B** 111

Bouchiat, M.A. et al. (1982): Phys. Lett. **117B** 358

Bouchiat, C., C.A.Piketty (1983): LPTENS preprint 83/18

Bouchiat, M.A. et al. (1983): Phys. Lett. **128B** 73; Nucl. Phys. **B221** (1983) 68

Bouchiat, M.A. et al. (1984): Phys. Lett. **134B** 463

Branco, G., J.M. Frére, J.M. Gerard (1983): Nucl. Phys. **B221** 317

Brandelik, R. et al. (1978): Phys. Lett. **73B** 162

Brandelik, R. et al. (1979): Phys. Lett. **83B** 261

Brandelik, R. et al. (1980): Z. f. Phys. **C4** 87

Brandelik, R. et al. (1982): Phys. Lett. **113B** 499

Braunschweig, W. (1988): Proceedings of the 24th International Conference on High Energy Physics, Munich, 1988

Braunschweig, W. et al.(1988): Contribution to the 24th International Conference on High Energy Physics, Munich, 1988

Brown, R.W., R. Decker, E.A. Paschos (1984): Phys. Rev. Lett. **52** 1192

Buchmüller, W., S.H.H. Tye (1980): Fermilab - PUB - 80/94 -

Bucksbaum, P., E. Commins, L. Hunter (1981): Phys.Rev. Lett. **46** 640

Buras, A.J., K.J.F. Gaemers (1978): Nucl. Phys. **B132** 249

Buras, A.J. (1981): Phys. Rev. Lett. **46** 1354

Buras, A.J. (1984): Proceedings of the Workshop on the Future of Medium Energy Physics in Europe, Freiburg

Buras, A.J. (1988): MPI-PAE/PTh 1/88, to appear in "CP Violation", C. Jarlskog (Editor), World Scientific (1988)

Burchat, P.R. et al. (1985): Phys. Rev. Lett. **54** 2489

Burger, J.D. (1982): Proceedings of the 21st International Conference on High Energy Physics, Paris 1982, p. C3-63

Cabibbo, N. (1963): Phys. Rev. Lett. **10** 531

Cahn, R.N., F.J. Gilman (1978): Phys. Rev. **D17** 1313

Carmony, D. D. et al. (1982): Phys. Rev. **D26** 2965

Carr, J. et al. (1983): Phys. Rev. Lett. **51** 627

Celmaster, W., R.J. Gonsalves (1980): Phys. Rev. Lett. **44** 560

Chadwick, K. et al. (1983): Phys. Rev. **D27** 473

Chen, H.H., B.W. Lee (1972): Phys. Rev. **D5** 1874

Cheng, T.P and F. Li (1977): Phys. Rev. Lett. **38** 381

Chertyrkin, K.G. et al. (1979): Phys. Lett. **85B** 277

Chrin, J. (1987): Z. Phys. **C36** 163

Clark, A.G. (1986): to appear in the Proceedings of the 12th International
    Conference on Neutrino Physics and Astrophysics, Sendai, Japan

Cnops, A.M. et al. (1978): Phys. Rev. Lett. **41** 357

Colas, P. et al. (1988) CERN-EP/88-16

Collins, J.C., A.J. Mac Farlane (1974): Phys. Rev. **D10** 1201

Commins, E.D., P.H. Bucksbaum (1980): Ann. Rev. Nucl. Part. Sci. **30** 1

Commins, E.D. (1981): Atomic Physics, Vol. 7, p. 121, eds. D. Kleppner,
    F.M. Pipkin, Plenum, New York (1981):

Criegee, L., G. Knies (1982): Phys. Rep. **83** 153

D'Agostini, G. et al. (1988): Contribution to the 24th International Confer-
    ence on High Energy Physics, Munich, 1988

Dakin, J., G. Feldman (1975): Phys. Rev. **D 8** 1862

Davier, M. (1982): Proceedings of the 21st International Conference on High
    Energy Physics, Paris 1982, p. C3-471

Deden, H. et al. (1979): Nucl. Phys. **B149** 1

Deden, H. et al. (1979a): Proceedings of the International Conference on Neu-
    trino Physics '79, Bergen, Vol. 2, p. 397

Delfino, M.C. (1985): Thesis, Univ. of Wisconsin, unpublished

Derrick, M. et al. (1985): Argonne Preprint ANL-HEP- PR-85-114

Desman, E. (1973): Phys. Rev. **D7** 2755

Dine, M., J. Sapierstein (1979): Phys. Rev. Lett. **43** 668

Diwan, M. (1988): Proceedings of the 13th International Conference on Neu-
    trino Physics and Astrophysics, Neutrino '88, Boston

Dorfan, J.M. et al. (1979): Phys. Rev. Lett. **43** 1555 (1981); Phys. Rev. Lett.
    46 215

Dowell, J. D. (1986): Proceedings of the 6th Topical Workshop on Proton-
    Antiproton Collider Physics, Aachen, 1986, p.419

Drell, P.S., E.D. Commins: Univ. of Calif. Preprint, Berkeley, May 1984

Duke, D. W., R.G.Roberts (1985): Phys. Rep. **120** 275

Dydak, F. (1983): Proceedings of the 1983 International Symposium on Lep-
    ton and Photon Interactions at High Energies, Cornell University, p.634

Ellis, J. (1977): Weak and Electromagnetic Interactions at High Energy, North-
    Holland, Amsterdam 1977, p. 5

Ellis, J. (1983): Proceedings of the 1983 International Symposium on Lepton
    and Photon Interactions at High Energies, Cornell University, p. 439

Ellis, J., D.V. Nanopoulos, S. Rudaz (1982): Nucl. Phys. **B202** 43

Emmons, T.P. et al. (1983): Phys. Rev. Lett. **51** 2089

Englert, F., R. Brout (1964): Phys. Rev. Lett. **13** 321

Entenberg, A. et al. (1978): Phys. Rev. Lett. **42** 1198

Errede, S. (1988): Proceedings of the 24th International Conference on High Energy Physics, Munich, 1988

Erriquez, Q. et al. (1980): Nucl. Phys. **B176** 37

Faissner, H. et al. (1978): Phys. Rev. Lett. **41** 213

Faissner, H. et al. (1983): Phys. Lett. **125B** 230

Fajfer, S. and R.J. Oakes (1984): Phys. Rev. **D30** 1585

Fayet, P., S. Ferrara (1977): Phys. Rep. **32C** 249

Feldman, G.J. (1978): SLAC - PUB - 2230

Feldman, G.J. et al. (1982): Phys. Rev. Lett. **48** 66

Fermi, E. (1983): 1934, Nuovo Cim. **11** 1

Fernandez, E. et al. (1983): SLAC-PUB-3133, July 1983

Fernandez, E. et al. (1983a): Phys. Rev. Lett. **50** 2054

Fernandez, E. et al. (1983b): Phys. Rev. Lett. **51** 1022

Fernandez, E. et al. (1985): Phys. Rev. Lett. **54** 1624

Fernandez, E. et al. (1985a): Phys. Rev. Lett. **54** 1620

Fernandez, E. et al. (1985b): Phys. Rev. **D31** 1537

Feynman, R.P., M. Gell-Mann (1958): Phys. Rev. **109** 193

Feynman, R.P. (1969): Proceedings of the Third International Conference on High Energy Collisions, Stony Brook

Field, R.D., R.P. Feynman (1978): Nucl. Phys. **B136** 1

Fiorini, E. (1983): Proceedings of the 1983 International Symposium on Lepton and Photon Interactions at High Energies, Cornell University, p. 405

Flügge, G. (1979): Z. Phys. **C1** 121

Fogli, G.L. (1985): Nucl. Phys. **B260** 593; Phys. Lett. **158B** 66

Ford, W.T. (1982): Proceedings of the International Conference on Instrumentation, SLAC Report No SLAC-250 (unpublished)

Ford, W.T. et al. (1982a): Phys. Rev. Lett. **49** 106

Ford, W.T. et al. (1986): Contribution to the 23rd International Conference on High Energy Physics, Berkeley, California

Fortson, E.N., L. Wilets (1980): Adv. Atom. Molec. Phys. **1649** 319

Fortson, E.N., L. L. Lewis (1984): Phys. Rep. **113** 289

Friedman, J.I., V.L. Telegdi (1957): Phys. Rev. **105** 1681

Fritzsch, H., M. Gell-Mann, H. Leutwyler (1973): Phys. Lett. **47B** 365

Fritzsch, H., P. Minkowski (1975): Ann. Phys. **93** 193

Fritzsch, H. (1979): Z. Phys. **C 1** 321

Fritzsch, H., D. Schildknecht, R. Kögerler (1982): Phys. Lett. **114B** 157

Fritzsch, H. (1983): MPI preprint MPI-PAE/Pth 47/83; 76/83

Galik, R.S. (1985): Proceedings of the International Europhysics Conference on High Energy Physics, Bari, p. 219

Gan, K.K. et al. (1985): Phys. Lett. **153B** 116

Garwin, R. et al. (1957): Phys. Rev. **105** 1415

Gelfand, Y.A., E.P. Likhtman (1971): Pis'ma Zh. Eksp. Teor. Fiz. **13** 323

Georgi, H., S.L. Glashow (1974): Phys. Rev. Lett. **32** 438

Georgi, H. et al. (1974): Phys. Rev. Lett. **33** 451

Georgi, H., H.D. Politzer (1976): Phys. Rev. **D14** 1829

Geweniger, C. et al. (1979): Neutrino- 79, Bergen, Vol. 2, p. 392

Gilbert, S. L. et al. (1985): Phys. Rev. Lett. **55** 2680

Gilman, F.J., D. H. Miller (1978): Phys. Rev. **D17** 1846

Gilman, F.J., T. Tsao (1979): Phys. Rev. **D19** 790

Gilman, F.J., S.H. Rhie (1985): Phys. Rev. **D31** 1066

Glashow, S.L. (1961): Nucl. Phys. **22** 579

Glashow, S.L., J. Iliopoulos, L. Maiani (1970): Phys. Rev. **D2** 1285

Goggi, G. (1979): Proceeding of the LEP Summer Study, CERN 79-01, p. 483

Gorishny, S. G. et al. (1988): JINR E2-88-254

Grabosch, H. J. et al. (1986): Z. Phys. **C31** 203

Greenshaw, T. et al. (1988): Contribution to the 24th International Conference on High Energy Physics, Munich, 1988

Gunion, J. (1988): Proceedings of the 24th International Conference on High Energy Physics, Munich, 1988

Guralnik, G.S. et al. (1964): Phys. Rev. Lett. **13** 585

Haguenauer, M. (1974): Proc. of the XVII International Conference on High Energy Physics, London 1974, p. IV-37

Hall, L. (1981): Nucl. Phys. **B178** 75

Harris, F. A. et al. (1977): Phys. Rev. Lett. **39** 437

Hasert, F.J. et al. (1973): Phys. Rev. Lett. **46B** 138

Heinzelmann, G. (1982): Proceedings of the 21st International Conference on High Energy Physics, Paris 1982, p. C3-59

Heisterberg, R.H. et al. (1980): Phys. Rev. Lett. **44** 635

Herb, S.W. et al. (1977): Phys. Rev. Lett. **39** 252

Higgs, P.W. (1964): Phys. Rev. Lett. **13** 508

Hitlin, D. (1987): Proceedings of the 1987 International Symposium on Lepton and Photon Interactions at High Energies, Hamburg, p. 179

Hollister, J.H. et al. (1981): Phys. Rev. Lett. **46** 643

Hung, P.Q. (1977): Phys. Lett. **69B** 216

Hung, P.Q. (1978): Phys. Rev. **D17** 1893

Hung, P.Q., J.J. Sakurai (1977): Phys. Lett. **69B** 323

Hung, P.Q., J.J. Sakurai (1981): Ann. Nucl. Part. Sci. **31** 375

Isiksal, E. et al. (1984): Phys. Rev. Lett. **52** 1096

Jacob, M. (1958): Nuovo Cim. **9** 826

Jacob, M., G. C. Wick (1959): Ann. Phys. **7** 404

Jadach, S. (1983): private communication

Jadach, S. (1985): Act. Phys. Pol. **B16** 1007

Jaros, J.A. et al. (1978): Phys. Rev. Lett. **40** 1120

Jaros, J.A. et al. (1983): Phys. Rev. Lett. **51** 955

Jaros, J.A. (1984): Proceedings of the International Conference on Physics in Collision IV, Santa Cruz

Jegerlehner, F. (1986): Z. Phys. **C32** 195, 435

Jenni, P. (1987): Proceedings of the 1987 International Symposium on Lepton and Photon Interactions at High Energies, Hamburg, p. 341

Jersak, J. et al. (1981): Phys. Lett. **98B** 363

Jones, E. et al. (1983): Design Study of an Antiproton Collector (ACOL), CERN 83-10, Geneva

Jones, G. T. et al. (1986): Phys. Lett. **178B** 329

Jonker, M. et al. (1980): Phys. Lett. **93B** 203

Jonker, M. et al. (1981): Phys. Lett. **99B** 265, (E) **103B** 469

Jonker, M. et al. (1982): Nucl. Instr. Methods **200** 183

Kafka, T. et al. (1982): Phys. Rev. Lett. **48** 910

Kagan, H. et al. (1983): Rochester Preprint UR 854 (1983):

Kayser, B. et al. (1979): Phys. Rev. **D20** 87

Kalmus, G. (1982): Proceedings of the 21th International Conference on High Energy Physics, Paris, p. C3-431

Kamae, T. (1988): Proceedings of the 24th International Conference on High Energy Physics, Munich, 1988

Kane, G.L., M.E. Peskin (1982): Nucl. Phys. **B195** 29

Kawamoto, N., A. I. Sanda (1978): Phys. Lett. **76B** 446

Kibble, T.W.B. (1967): Phys. Rev. **155** 1554

Kim, J.E. (1977): Phys. Rev. **D16** 172

Kim, J.E. et al. (1981): Rev. Mod. Phys. **53** 211

Kirkby, J. (1979): Proc. 1979 Int. Symp. Lepton Photon Interact. High Energies, FNAL, p. 107

Kirkby, J. (1982): Proceedings of the 21st International Conference on High Energy Physics, Paris 1982, p. C3-45

Kiesling, C. (1982): Proceedings of the XIII International Symposium on Multiparticle Dynamics, Volendam, Netherlands, p. 577

Kiesling, C. (1985): Proceedings of the XXth Rencontre de Moriond, p. 65 and MPI-PAE/Exp. El. 155

Kiesling, C. (1988): τ Physics, in High Energy e^+e^- Physics, edited by A.ALi, P.Söding, World Scientific, Singapore (1988)

Klein, M. (1985): Fortschr. Phys. **33** 375

Klem, D.E. et al. (1984): Phys. Rev. Lett. **53** 1873

Kluttig, H. et al. (1977): Phys. Lett. **71B** 446

Kobayashi, M., T. Maskawa (1972): Progr. Theor. Phys. **49** 282

Kobayashi, M., T. Maskawa (1973): Progr. Theor. Phys. **49** 652

Komamiya, S. (1985): Proceedings of the 1985 International Symposium on Lepton and Photon Interactions at High Energies, p. 612

Krenz, W. et al. (1978): Nucl. Phys. **B135** 45

Krenz, W. (1985): PITHA 84-42, Aachen, March 1985

Kuroda, M., D. Schildknecht (1982): Phys. Rev. **D26** 3167

La Rue, G.S., J.D. Phillips, W.M. Fairbank (1981): Phys. Rev. Lett. **46** 976

Lee, B.W. (1972a): Phys. Rev. **D5** 823

Lee, B.W. (1972b): Proceedings of the XVI International Conference on High Energy Physics, edited by J.D. Jackson and A. Roberts (Fermi National Accelerator Cents, Batavia, Illinois), Vol IV, p. 266

Lee, B.W., R. Zinn-Justin (1972a): Phys. Rev. **D5** 3121

Lee, B.W., R. Zinn-Justin (1972b): Phys. Rev. **D5** 3137

Lee, B.W., R. Zinn-Justin (1972c): Phys. Rev. **D5** 3155

Lee, T.D., C.N. Yang (1956): Phys. Rev. **104** 254

Lepage, G.P. (1983): Proceedings of the 1983 International Symposium on Lepton and Photon Interactions at High Energies, Cornell University, p. 565

Levi, N.E. et al. (1983): Phys. Rev. Lett. **51** 1941

Lewis, L.L. et al. (1977): Phys. Rev. Lett. **39** 795

Liede, I. et al. (1978): Nucl. Phys. **B146** 157

Llewellyn Smith, C.H. (1983): Proc. Workshop on SPS Fixed-Target Physics in the Years 1984-1989, CERN 83-02, Vol. II, p. 180

Llewellyn Smith, C.H., G.G. Ross, J.F. Wheater (1981): Nucl. Phys. **B177** 263

Llewellyn Smith, C.H., J.F, Wheater (1981): Oxford University preprint 68/81

Lockyer, N.S. et al. (1983): SLAC.- PUB - 3245, October 1983

Lockyer, N.S. et al. (1983a): Phys. Rev. Lett. **51** 1316

Lüth, V. (1985): Proceedings of the International Conference on Physics in Collision V, Autun, France, 1985

Ma, E. (1983): Proceedings of the International Europhysics Conference on High Energy Physics, Brighton (1983) p. 259

Madaras, R.J. (1984): Proceedings of the XIXth Rencontre de Moriond, Vol 2

Marciano, W.J., H. Pagels (1978): Phys. Rev. **36C** 137

Marciano, W.J. (1979): Phys. Rev. **D20** 274

Marciano, W.J., A. Sirlin (1980): Phys. Rev. **D22** 2695

Marciano, W.J., A. Sirlin (1981): Phys. Rev. Lett. **46** 163

Marciano, W.J. (1983): DESY 83-111, Proceedings of the 1983 International Symposium on Lepton and Photon Interactions at High Energies, Cornell University, p. 80

Marciano, W.J., A. Sirlin (1983): Phys. Rev. **D27** 52

Marciano, W.J., A. Sirlin (1984): Phys. Rev. **D29** 945

Marciano, W.J. (1985): Proceedings of the Aspen Winter Physics Conference, Jan. 1985

Marriner, J. et al. (1977): preprint LBL-6438

Martenssen, A. M. et al. (1981): Phys. Rev. **A24** 308

Martenssen-Pendrill, A. M. (1985): J. Physique **46** 1948

Maruyama, T (1986): Proceedings of the 23rd International Conference on High Energy Physics, Berkeley, California, p. 975

Maxeiner, C. (1985): DESY Internal Report PLUTO-85-03

Maxwell, C.J., M.J.Teper (1981): Z. f. Phys. **C7** 253

Merritt, F.S. et al. (1978): Phys. Rev. **D17** 2199

Mess, K. et al. (1979): Neutrino- 79, Bergen, Vol. 2, p. 371

Meyer, H. (1986): Proceedings of the International Symposium on Weak and Electromagnetic Interactions in Nuclei, Heidelberg, 1986, p. 846

Mills, G.B. et al. (1984): Phys. Rev. Lett. **52** 1944

Mills, G.B. et al. (1985): Phys. Rev. Lett. **54** 624

Mo, L.W., Y.S.Tsai (1978): Rev. Mod. Phys. **41** 204

Mohapatra, R.N., G. Senjanovich (1980): Phys. Rev. Lett. **40** 912

Mohapatra, R.N., G. Senjanovich (1981): Phys. Rev. **D23** 165

Moreels, J. et al. (1984): Phys. Lett. **138B** 230

Mueller, A. (1981): Phys. Rep. **73C** 237

Murtagh, M. J. (1984): Proceedings of the 11th International Conference on Neutrino Physics and Astrophysics, Dortmund (1984) p. 290

Myatt, G. (1982): Rep. Prog. Phys. **45** 1

Nelson, M.E. et al. (1983): Phys. Rev. Lett. **50** 1542

Panman, J. (1984): Proceedings of the XIth International Conference on Neutrino Physics and Astrophysics, Dortmund, p. 741

Particle Data Group, M. Roos, et al. (1982): Phys. Lett. **111B** 1

Paschos, E.A., L. Wolfenstein (1973): Phys. Rev. **D7** 91

Paschos, E.A., M. Wirbel (1982): Nucl. Phys. **B194** 189

Pasierb, E. et al. (1979): Phys. Rev. Lett. **43** 96

Passarino, G., M. Veltman (1979): Nucl. Phys. **B160** 151

Pati, J.C., A. Salam (1974): Phys. Rev. **D10** 275

Peccei, R.D. (1984): Phys. Lett. **136B** 121

Perl, M. et al. (1975): Phys. Rev. Lett. **35** 1489

Perl, M. (1980): Ann. Rev. Nucl. Part. Sci. **30** 299

Peskin, M. (1980): Proceeding of the International Symposium on Lepton and Photon Interactions at High Energies, Bonn, p. 880 .

Pham, T.N., C. Roiesnel, T.N. Truong (1978): Phys. Lett. **78B** 623

Piccolo, M. (1983): Frascati preprint LNF - 83/64 (P)

Piketty, C.A. (1984): Proceedings of the XIth International Conference on Neutrino Physics and Astrophysics, Dortmund

Politzer, H.D. (1974): Phys. Rep. **14C** 129

Politzer, H.D. (1979): Phys. Lett. **84B** 524

Prescott, C.Y. et al. (1978): Phys. Lett. **77B** 347

Prescott, C.Y. et al. (1979): Phys. Lett. **84B** 524

Pullia, A. et al. (1979): Proceedings of the International Conference on Neutrino Physics '79, Bergen, p. 230

Pullia, A. (1984): Riv. Nuovo Cim. **7** 1

Rein, D. and L.M. Sehgal (1983): Nucl. Phys. **B223** 29

Reines, F. et al. (1976): Phys. Rev. Lett. **37** 315

Reutens, P. G. et al. (1985): Phys. Lett. 52 B 404

Ritson, D. M. (1986): Proceedings of the 23rd International Conference on High Energy Physics, Berkeley, California, p. 809

Roe, B. (1979): Proceedings of the International Conference on Neutrino Physics '79, Bergen, Vol. 2, p. 592

Ross, G. G. (1987): Proceedings of the 1987 International Symposium on Lepton and Photon Interactions at High Energies, Hamburg, p. 743

Rubbia, C., P. Mc Intyre, D. Cline (1976): Proceeding International Neutrino Conference Aachen 1976, p. 683

Rubbia, C. (1985): Proceedings of the 1985 International Symposium on Lepton and Photon Interactions at High Energies, p. 242

Ruckstuhl, W. et al. (1986): Phys. Rev. Lett. **56** 2132

Salam, A. (1968): Elementary Particle Theory, ed. N. Swartholm (Stockholm: Almquist and Wiksell) p. 367

Salam, A., J.C.Ward (1964): Phys. Rev. Lett. **13** 168

Sarantakos, S. et al. (1983): Nucl. Phys. **B217** 84

Schaeffer, R. et al. (1987): Nature **300** 142

Schildknecht, D. (1983): MPI preprint MPI-PAE/Pth 23/83

Schindler, R.H. et al. (1981): Phys. Rev. **D24** 78

Schmidke, W. B. et al. (1986): Phys. Rev. Lett. **57** 527

Schwinger, J. (1957): Ann. Phys. **2** 407

Sciulli, F. (1979): Prog. in Nucl. and Part. Phys. **2** 41

Sciulli, F. (1985): Proceeding of the International Symposium on Lepton and Photon Interactions at High Energies, Kyoto, p. 8

Sehgal, L.M. (1975): Nucl. Phys. **B90** 471

Sehgal, L.M. (1977): Phys. Lett. **71B** 99

Sehgal, L.M. (1980): PITHA 80/17

Sirlin, A. (1980): Phys. Rev. **D22** 971

Sirlin, A. (1984): Phys. Rev. **D29** 89

Spencer, L. J. et al. (1981): Phys. Rev. Lett. **47** 771

Steffen, P. (1984): Proceedings of the International Conference on High Energy Physics, Leipzig

Steigman, G. et al. (1986): Phys. Lett. **176B** 33

Stirling, W. J. (1987): Proceedings of the 1987 International Symposium on Lepton and Photon Interactions at High Energies, Hamburg, p. 715

Sudarshan, E.C.G., R. Marshak (1958): Phys. Rev. **109** 1860

Suzuki, M. (1977): Phys. Lett. **71B** 139

Talaga, R. (1988): Proceedings of the 24th International Conference on High Energy Physics, Munich, 1988

Thacker, H. B., J. J. Sakurai (1971): Phys. Lett. **36B** 103

The Staff of the CERN proton-antiproton project (1981): Phys. Lett. **107B** 306

t'Hooft, G. (1971): Phys. Lett. **37B** 195

t'Hooft, G. (1971a): Nucl. Phys. **B33** 173

t'Hooft, G. (1971b): Nucl. Phys. **B35** 167

t'Hooft, G., M. Veltman (1972a): Nucl. Phys. **B44** 189

t'Hooft, G., M. Veltman (1972b): Nucl. Phys. **B50** 318

Thorndike, E.H. (1985): Proceeding of the International Symposium on Lepton and Photon Interactions at High Energies, Kyoto (1985):

Trilling, G.H. (1982): Proceedings of the 21st International Conference on High Energy Physics, Paris 1982, p. C3-57

Tsai, Y.S. (1971): Phys. Rev. **D4** 2821

Tsai, Y.S. (1980): SLAC - PUB - 2403

Unno, Y. (1988): Proceedings of the 24th International Conference on High Energy Physics, Munich, 1988

Veltman, M. (1977): Nucl. Phys. **B123** 89

Veltman, M. (1977a): Phys. Lett. **B70** 253

Van der Meer, S. (1972): CERN/ISR-PO / 72-31

Volkov, D., V.P. Akulov (1973): Phys. Lett. **46B** 109

Wanderer, P. et al. (1978): Phys. Rev. **D17** 1679

Weinberg, S. (1967): Phys. Rev. Lett. **19** 1264; Phys. Rev. Lett. **27** 1688

Weinberg, S. (1976): Phys. Rev. Lett. **36** 294

Weinstein, R. (1982): Particles and Fields, edited by W.E. Caswell and G.A. Shaw, AIP Proceedings No 98, American Institute of Physics, New York (1982), p. 126

Weiss, J. M. (1984): ANL-HEP-CP-84-59, preprint and contribution to the Leipzig Conference 1984

Wess, J., B. Zumino (1974): Nucl. Phys. **B70** 39

Wetzel, W. (1983): Nucl. Phys. **B227** 1 and Heidelberg Preprint May 1983

Wheater, J.F. (1981): Phys. Lett. **105B** 483

Wheater, J.F., C.H. Llewellyn Smith (1982): Nucl. Phys. **B208** 27; **B226** (1983) 547 (E)

Wolfenstein, L. (1978): Nucl. Phys. **B146** 477

Wu, C.S.E. et al. (1957): Phys. Rev. **105** 1413

Wu, Sau Lan (1987): Proceedings of the 1987 International Symposium on Lepton and Photon Interactions at High Energies, Hamburg

Yamada, S. (1983): Proceedings of the International Lepton - Photon Symposium, Cornell University, Ithaca , p.525

Yamamoto, H. et al. (1985): Phys. Rev. Lett. **54** 522

Yang, C.N., R. Mills (1954): Phys. Rev. **96** 191

Yelton, J.M. et al. (1982): Phys. Rev. Lett. **49** 430

Zacek, V. (1988): Proceedings of the the 24th International Conference on High Energy Physics, Munich, 1988

Subject Index

Jacobian peak   164
JADE experiment   42
jet chamber   42
jet charge   91

K factor   164
K_{e3} decays   140

Lagrangian
  charged current   4
  neutral current   12
leading particle   91
left-handed coupling   11
lepton tagging
  of heavy quarks   94–97
lepton-nucleon scattering   176
leptonic decays
  of the τ 63
  of the $W^{\pm}$ 167–168
  of the Z^0 170
lifetime
  of the proton   187
  of the τ 66–67
liquid argon calorimeter   43
longitudinal polarisation   75
loop corrections   38
low energy (local) limit   108, 146
luminosity   40

MAC experiment   42
magnetic dipole transition   152
MARK II experiment   42
MARK J experiment   42
Michel parameter   67
minimal coupling   6
Monte Carlo technique   36
μ lepton   49–57
μC experiment   157
muon spectrometer   106
$\overline{MS}$ scheme   88

narrow band beam   139
NC reactions   105, 107
neutral current
  flavour-changing   102–103
  interactions   105

ν_e beam   118
neutrino flux   104
November Revolution   13

one loop correction   172, 174
optical rotation   152

parametrisation
  of weak propagator   21
parity conservation   4
parity violation   4, 9, 19, 75, 146, 152
  maximal   9
Paschos-Wolfenstein relation   139
PCAC hypothesis   60
PEP e^+e^- storage ring   39–40
PETRA e^+e^- storage ring   40
π^0 reconstruction   64–65
PLUTO experiment   42
polarisation
  asymmetry   28–29, 77–78, 147
  average   27
  longitudinal   30
  of final state fermions   26
  of the τ 77
polarised electrons   146
propagator   21
  function   20, 33
ψ production   178

Q value   94
quantum chromodynamics (QCD)   1, 84
quantum electrodynamics (QED)   5
quark model   25
quark parton model   84, 119–120
quasi-elastic scattering   110–111

radiative correction
  to the boson masses   15, 172
radiative corrections
  in e^+e^- reactions   34–39
reactor neutrino experiment   113
reduced QED correction   36
renormalisability   181
renormalisation scheme
  $\overline{MS}$ 88
  on-shell   172

211

Springer Tracts in Modern Physics

Further volumes of the *Springer Tracts in Modern Physics* dealing with *nuclear* and *particle* physics are

Volumes 36, 39, 41, 45, 49, 55, 57, 59, 63, 71, 79, 83, 86, 89, 90, 100–102, 105, 108.

In this very year Vol. 108 became rather popular:
It is a Festschrift for **Jack Steinberger** who was awarded the Nobel prize in physics 1988 together with **Leo Lederman** and **Melvin Schwartz** who presented contributions to this volume which also contains still further articles by Nobel prize winners of previous years.

Particles and Detectors

Festschrift for Jack Steinberger
Editors: K. Kleinknecht, T.D. Lee

This volume contains a collection of 20 reviews on elementary particle physics by well-known scientists, including by now (i.e., 1988) five Nobel prize-winners, and is dedicated to Jack Steinberger on his 65th birthday. The subjects of the contributions range from theoretical particle physics and phenomenology to experimental results on electroweak interactions, neutrino physics, and particle detectors.

1986. 91 figs. X, 291 pp.
(Springer Tracts in Modern Physics, Vol. 108)
Hard cover DM 99,–. ISBN 3-540-16265-8

Founded in 1922 under the name "Ergebnisse der exakten Naturwissenschaften" this series received its present name, *Springer Tracts in Modern Physics*, in 1965 and for the last quarter-century, since 1963, has been edited with distinction by Professor G. Höhler and Professor E.A. Niekisch.

On the occasion of the 25th anniversary of the Springer Tracts we come up with a

SPECIAL OFFER

to give you a chance to fill gaps in your collection. While stock lasts we offer you at half the former list price **Volumes 36–89** (for a complete list see overleaf; the prices are subject to change without notice).

Vol. 36: 1964. 57 figs. IV,242 pages. (Special ed. s. Lichtenberg, Meson and Baryon Spectroscopy) 530g Hard DM 29,50
(3-540-03215-0)

Vol. 39: **Electron and Photon Interactions at High Energies.** 1965. 122 figs., 10 tables. VIII,168 pages. 420g Hard DM 29,–
(3-540-03406-4)

Vol. 41: **Wildermuth, K.; McClure, W.:** Cluster Representations of Nuclei. 1966. 32 figs. VI,172 pages. 430g Hard DM 32,–
(3-540-03670-9)

Vol. 45: **Collins, P. D. B.; Squires, E. J.:** Regge Poles in Particle Physics. 1968. 90 figs. VIII,292 pages. 610g Hard DM 49,50
(3-540-04339-X)

Vol. 48: **Grosse, P.:** Die Festkörpereigenschaften von Tellur. 1969. 103 Abb. IV,208 Seiten. 455g Geb DM 32,–   (3-540-04711-5)

Vol. 55: **Low Energy Hadron Interactions.** Invited Papers presented at the Ruhestein-Meeting, May 1970. 1970. 49 figs. 33 tables. V,290 pages. 575g Hard DM 49,50
(3-540-05250-X)

Vol. 57: **Strong Interaction Physics.** Heidelberg-Karlsruhe. International Summer Institute in Theoretical Physics (1970) 1971. 65 figs. V,270 pages. 580g Hard DM 48,–
(3-540-05252-6)

Vol. 59: **Symposium on Meson-, Photo-, and Electroproduction at Low and Intermediate Energies.** Bonn, September 21 - 26, 1970. 1971. 154 figs. III,222 pages. 500g Hard DM 48,–   (3-540-05494-4)

Vol. 63: **Photon-Hadron Interactions II.** International Summer Institute in Theoretical Physics, Desy, July 12-24, 1971. 1972. 97 figs. V,189 pages. 450g Hard DM 43,–
(3-540-05813-3)

Vol. 64: **Springer, T.:** Quasielastic Neutron Scattering for the Investigation of Diffusive Motions in Solids and Liquids. 1972. 36 figs. III,100 pages. 310g Hard DM 21,–
(3-540-05808-7)

Vol. 66: **Quantum Statistics in Optics and Solid-State Physics.** 1973. 30 figs. IV,173 pages. 430g Hard DM 46,80   (3-540-06189-4)

Vol. 67: **Ferrara, S.; Gatto, R.; Grillo, A. F.:** Conformal Algebra in Space-Time and Operator Product Expansion. 1973. IV,69 pages. 250g Hard DM 22,80   (3-540-06216-5)

Vol. 68: **Solid-State Physics.** 1973. 77 figs. 48 tab. III,205 pages. 445g Hard DM 52,80
(3-540-06341-2)

Vol. 70: **Agarwal, G. S.:** Quantum Optics. 1974. II,135 pages. 340g Hard DM 46,20
(3-540-06630-6)

Vol. 71: **Nuclear Physics.** 1974. 116 figs. III,245 pages. 510g Hard DM 58,80
(3-540-06641-1)

Vol. 72: **Langbein, D.:** Theory of Van der Waals Attraction. 1974. 32 figs. II,145 pages. 385g Hard DM 46,80   (3-540-06742-6)

Vol. 74: **Solid-State Physics.** 1974. 75 figs. III,153 pages. 340g Hard DM 46,80
(3-540-06946-1)

Vol. 79: **Elementary Particle Physics.** 1976. 37 figs. VI,145 pages. 410g Hard DM 35,–
(3-540-07778-2)

Vol. 81: **Leibfried, G.; Breuer, N.:** Point Defects in Metals I. Introduction to the Theory. 1978. 138 figs., 22 tab. XIV,342 pages. 710g Hard DM 44,50   (3-540-08375-8)

Vol. 82: **Bendow, B.; Lengeler, B.:** Electronic Structure of Noble Metals and Polariton-Mediated Light Scattering. 1978. 42 figs., 20 tab. VI,114 pages. 360g Hard DM 26,40
(3-540-08814-8)

Vol. 83: **Amaldi, E.; Fubini, S.; Furlan, G.:** Pion-Electroproduction. Electroproduction at Low Energy and Hadron Form Factors. 1979. 47 figs., 13 tab. VIII,162 pages. 425g Hard DM 32,–   (3-540-08998-5)

Vol. 84: **Olson, C. L.; Schumacher, U.:** Collective Ion Acceleration. 1979. 63 figs., 9 tab. VII,231 pages. 525g Hard DM 34,–
(3-540-09066-5)

Vol. 86: **Wiik, B. H.; Wolf, G.:** Electron-Positron Interactions. 1979. 238 figs., 43 tab. IX,262 pages. 625g Hard DM 36,–
(3-540-09604-3)

Vol. 88: **Raether, H.:** Excitation of Plasmons and Interband Transitions by Electrons. 1980. 121 figs., 17 tab. VIII,196 pages. 485g Hard DM 29,50   (3-540-09677-9)

Vol. 89: **Cannata, F.; Überall, H.:** Giant Resonance Phenomena in Intermediate-Energy Nuclear Reactions. 1980. 42 figs., 6 tab. VIII,112 pages. 380g Hard DM 26,–
(3-540-10105-5)